Natural Gas
Measurement
and Control

Natural Gas Measurement and Control

A Guide for Operators and Engineers

Lohit Datta-Barua, Ph.D.

McGraw-Hill, Inc.

New York   St. Louis   San Francisco   Auckland   Bogotá
Caracas   Lisbon   London   Madrid
Mexico   Milan   Montreal   New Delhi   Paris
San Juan   São Paulo   Singapore
Sydney   Tokyo   Toronto

Library of Congress Cataloging-in-Publication Data

Datta-Barua, Lohit.
  Natural gas measurement and control : a guide for operators and
engineers /Lohit Datta-Barua.
      p.   cm.
  Includes bibliographical references and index.
  ISBN 0-07-015608-5
  1. Gas, Natural—Pipe lines. 2. Gas, Natural—Measurement.
3. Flow meters. I. Title.
  TN880.5.D34 1991
  665.7'44—dc20   91-27760
                                                                   CIP

1 2 3 4 5 6 7 8 9 0   DOC/DOC   9 7 6 5 4 3 2 1

ISBN   0-07-015608-5

*The sponsoring editor for this book was Gail Nalven, the editing
supervisor was Jim Halston, and the production supervisor was
Suzanne W. Babeuf. This book was set in Century Schoolbook by
McGraw-Hill's Professional Book Group composition unit.*

Printed and bound by R. R. Donnelley & Sons Company.

To my beloved wife, Manjula,
and daughters, Seebany and Indrany,
who provide courage, strength, hope, and happiness
in a world so complex and so turbulent.

Contents

Chapter 2. Gas Measurement: Operational and Accounting Considerations

Chapter 3. Measurement Engineering and Design

Preface

Natural Gas Measurement and Control: A Guide to Operators and Engineers is designed as a comprehensive but concise and practical guidebook for practicing engineers, engineering students, operating personnel, and gas accounting and auditing personnel. This author, in his years of involvement with oil and gas operation, felt the need for a single source book that provides practical guidance in every major aspect of gas measurement. These aspects consist of design, engineering, construction, operation, volume and energy determination, contractual obligations, physical accounting of gas, error calculation and volume adjustment, etc.

In some books, certain aspects are dealt with in detail in a highly technical manner impractical for day-to-day operating personnel. Certain accounting and operating ideas are not available anywhere. Such ideas have evolved and have been handed down from one person to another by measurement personnel in various gas companies. These ideas are based on experience, analytical investigation, or experimental findings.

Chapter 1 provides a general introduction to gas measurement and measurement field and office equipment. Above all, it provides the formulas relevant to gas measurement in one place.

Chapter 2 of this book provides many ideas for operating and accounting personnel. Tables showing errors for various conditions were determined at Conoco Inc. and Florida Gas Transmission Co. Figure 2.2, showing a graph for error due to a backward plate, was determined through tests run at United Gas Pipe Line's high-pressure meter proving facility. This author was directly or indirectly involved in many such studies. None of these data have been formally approved or verified by the industry or any agency. Therefore, these are provided for guidance only; however, they should give the reader an idea of the magnitude of a problem.

Chapter 3 should assist practicing engineers and engineering students in the design, specification, and construction of gas measurement and control facilities. Once again, such practical design information is not readily available in any book to the best of this author's knowledge.

Chapter 4 deals with gas quality and means to determine gas composition and its specific gravity and energy value. It introduces certain practical concepts of gas sampling and chromatographic analysis of natural gas. In addition, a simple method for calculating energy value (Btu) at actual moisture content and converting Btu values to a different pressure base is given.

Chapter 5 provides typical contractual language dealing with general terms and conditions relevant to gas measurement. This allows the operator or the engineer to understand contractual obligations and to design or to operate facilities accordingly.

In a nutshell, these five chapters should provide the reader with a complete, practical understanding of the complex gas measurement process.

Appendixes provide a metric conversion table relevant to natural gas and a glossary of measurement terms. The appendixes are intended to assist readers here in the United States and overseas in understanding the gas measurement business in an international environment. The technical concepts, the engineering and operational ideas, and errors and their corrections are universal in nature. Overall, this book should be of tremendous value to anybody in the natural gas business. A book of this nature and scope has been overdue and, I believe, fills a void. If readers agree, my efforts will have been worthwhile.

The author, the publisher, and the parties mentioned herein should not be held responsible for accuracy, reliability, or defensibility of any data, recommendations, suggestions, etc. The entire work is based on experience, knowledge, limited experiments, good judgment, and a commonsense approach. It is simply a single source of valuable practical information, generally not available in its entirety.

Lohit Datta-Barua

Acknowledgment

I must acknowledge my sincere appreciation to many individuals and organizations that directly or indirectly assisted in the preparation of this book. I am indebted to many coworkers and friends. I was fortunate to be working with people like W. G. Birkhead (retired) of United Gas Pipe Line Co., C. F. Drake (retired) of Natural Gas Pipeline Co. of America, C. L. Rousseau (retired past manager of training for United Gas Pipeline Co.). I also must acknowledge the opportunity to work and learn primarily at United Gas Pipe Line and then at the MidCon companies, namely United Texas Transmission Co. and Natural Gas Pipe Line Co. of America. In addition, many friends at other companies, namely Oklahoma Natural Gas Co., El Paso Natural Gas Co., Florida Gas Transmission Co., and Entex Inc., have helped me indirectly. The list goes on but I must mention a few more. They are G. B. Lynn of Oklahoma Natural Gas Co.; Rusty Woomer, Adam Durant, E. G. Kemmerer, and W. Boyt of United Gas Pipe Line; C. E. Allen of United Texas Transmission Co. (previously with United Gas Pipe Line and MidCon Services Co.); Oris A. Brehmer (retired, United Texas Transmission); and Winston Meyer of Entex.

I was also fortunate to be able to play a key role in a measurement training program at United Gas Pipe Line and a gas accounting course offered by Southern Gas Association. The preparation of training material and actual instruction gave me additional ideas as to the need in the industry. I have been also involved with the American Gas Association, the American Petroleum Institute fluid measurement standards committees, the Gulf Coast Gas measurement short course, and the International School of Hydrocarbon Measurement in various capacities. This industry involvement confirmed my conviction that a single source book for applied measurement is warranted. I am thankful to be involved with these organizations.

Without the assistance, direct or indirect, from these individuals and organizations, I do not think this book would have been possible.

Finally, I want to thank Wayne Burrel of Mercury Instrument Co. for encouraging me to do this work, Evonne Weathersby for word processing, Tom Harris for preparing the drawings for the manuscript, and Gail F. Nalven, senior editor of McGraw-Hill Publishing Co., for her effort in getting the manuscript into final shape.

Abbreviations

AGA	American Gas Association
AWG	American Wire Gauge
ANSI	American National Standards Institute
API	American Petroleum Institute
ASME	American Society of Mechanical Engineers
ASTM	American Society for Testing and Materials
BLM	Bureau of Land Management
DOT	Department of Transportation
FM	Factory Mutual
GPA	Gas Processors Association
GRI	Gas Research Institute
IEEE	Institute of Electrical and Electronics Engineers
ISA	Instrument Society of America
ISO	International Standards Organization
MAOP	maximum allowable operating pressure
MMS	Minerals Management Service
NACE	National Association of Corrosion Engineers
NEC	National Electric Code
NFPA	National Fire Protection Association
OCS	outer continental shelf
OSHA	Occupational Safety and Health Administration
RRC	Railroad Commission (of Texas)
SGA	Southern Gas Association
SMYS	specified minimum yield strength
UL	Underwriters Laboratory

1

Fundamentals of Gas Measurement and Flow Calculation

1.1 Introduction, Standards, and Fundamentals

General

Gas measurement is said to be the cash register of the gas industry. The efficient and accurate measurement of natural gas is of vital importance in the present-day world of energy management. Accounting for gas can be no better than the measurement of gas. Gas measurement is based on a combination of the laws of physics, chemistry, engineering, and accounting and is not an exact science. However, it is an applied science and, with changing technology and business environment, measurement technology is changing as well.

The open access concept, along with gas brokering and marketing, has resulted in a unique style of contracts in the United States of America. Multiple contracts through one physical measuring point have resulted in nomination and allocation challenges. These require faster information (daily, hourly) from the measuring point to make quick and prudent business decisions and to minimize imbalances.

These practices, and the trend to automated billing, have forced the gas industry to quickly move toward electronic measurement. Although primary gas flow devices for measurement are not expected to change in the near future, the secondary devices will increasingly be modern electronic equipment. These changes are also causing the gas industry to provide new training to engineering, measurement, chart processing, accounting, and auditing personnel to enable them to handle electronic equipment as well as information generated by such electronic measurement systems. In many companies, gas measurement is a function that is handled by a group other than accounting personnel. However, it is of vital importance to gas accounting. Accounting, as it applies generally to the gas industry, is based on conditions in contracts between the producer, or seller, and the buyer (or shipper and transporter) and on regulatory standards.

Units of measurement

All discussion shall be primarily in inch-pound (in · lb) units. However, for international use, metric units and a conversion table for the natural gas industry are provided in Appendix B. Generally, gas volumes are expressed in standard cubic feet (scf) or multiples thereof, such as thousand standard cubic feet (Mscf). In measuring gas, the unit is the cubic foot, or the quantity of gas that fills a dimensional cubic foot of space at a given pressure and temperature. Prior to 1970, the Mscf was an adequate unit of measure for the gas industry.

However, as gas has become more valuable, the unit of energy in British thermal units (Btu) or multiples thereof, such as million Btu (MMBtu), is becoming the unit of measurement. The British thermal unit is a measure of the heat value of a fuel. As applied in general practice, one Btu is the amount of heat required to raise the temperature of one pound of water by one degree Fahrenheit.

Gas measurement standards

Since gas volumes vary according to pressure and temperature, it is necessary to express volumes in accordance with some standard set of conditions. The common pressure bases are 14.73, 14.65, and 15.025 lb/in^2 absolute. Absolute pressure is the sum of gauge pressure and atmospheric pressure. The base temperature generally used throughout the gas industry is 60°F (520° absolute or 520°R). According to the kinetic theory of gases, there is zero molecular activity at the temperature of absolute zero. Absolute zero is −459.67°F; a value of 460° is used in gas measurement—i.e., 0°F = 460°R or 460° absolute. Absolute pressure and absolute temperature are used in all calculations.

AGA Report No. 3

This report, approved by the American Gas Association (as AGA Report No. 3), the American National Standards Institute and American Petroleum Institute (as ANSI/API 2530), and the Gas Processors Association (as GPA 8185-85), is the standard used in the measurement of natural gas by orifice meters in the United States. ISO 5167 (published by the International Standards Organization) is a standard used by some other countries. These provide the standards for construction and installation of orifice plates and associated fittings and the instructions for computing the flow of natural gas through orifice meters. Included also are the necessary tables of the basic factors for adjusting measurements of the temperature and pressure, such as specific gravity, supercompressibility factor, expansion factor, and Reynolds number factor.

Expansion factor is the density correction factor (density of flowing fluid changes as pressure changes in passing through orifice). *Reynolds number factor* is used to compensate for characteristic changes caused by velocity, viscosity, and density. Supercompressibility accounts for deviation from ideal gas laws.

Regulatory standards

In many states, standards have been established for reporting volumes for production tax purposes; in other states or federal areas, regulatory bodies have prescribed the standards to be used in determining volumes for reporting the production and disposition of gas. Examples are the Federal Energy Regulatory Commission (FERC) and the Texas Railroad Commission. Almost every country has some regulatory body establishing measurement standards for the oil and gas industry.

Contract standards

A contract between two parties entering into a gas transaction usually specifies the pressure and temperature base, the assumed atmospheric pressure, the Btu base, the starting and ending time of the month, and whether adjustments are to be made for various factors such as the Reynolds number factor, manometer factor, thermal expansion factor, gauge location factor, and expansion factor. Requirements for determining gas flowing temperature, gas gravities, and Btu content are also set out. Meter testing and measurement accuracy provisions are also included. With higher gas prices, measurement errors are usually corrected if they are over 1 percent.

Gas laws

Determination of the volume of gas at a given set of pressure and temperature conditions is fundamental to gas measurement. The relationships which tell the values of these quantities and how they are tied to each other are known as gas laws. They are different versions of the same fundamental law.

Boyle's law.   This law states that the volume of a gas varies inversely with the absolute pressure. If the temperature of a given quantity of gas is held constant, the volume of gas varies inversely with the absolute pressure.

Charles' law.   This law states that the volume of a gas varies directly with absolute temperature. Combining Boyle's law and Charles' law, we obtain the perfect gas law:

$$\frac{P_1 V_1}{T_1} = \frac{P_2 V_2}{T_2} = \text{constant} \tag{1.1}$$

However, a real gas such as natural gas deviates from the perfect gas law, and such deviation is accounted for by the compressibility factor.

1.2   Measurement Field Equipment

General

In order to measure gas accurately, many meters and related instruments are needed. Maintaining these meters and instruments and converting the readings and chart recordings into gas volumes are functions of the gas measurement personnel.

The complex job of measurement begins at the very moment gas leaves a gas well, pipeline delivery connection, or gas processing plant. At this point, measurement personnel take over the responsibilities of following the gas on its way from the source of supply to the market. Field measurement personnel include a chart changer and one or more levels of measurement technicians. Chart changers are responsible for placing, removing, and marking charts for processing, whereas measurement technicians are responsible for operation, maintenance, and calibration of equipment according to the terms of the contract, including witnessing of tests.

Field equipment

The physical function of gas measurement that is performed in the field is of vital importance. The accounting of gas can be no better than the measurement of the raw gas at its source. The accuracy of flowing conditions recorded by the measurement devices depends on the proper installation and subsequent functioning of the field equipment. The calibration and care of the measurement instruments in the field must be accomplished with a high degree of accuracy. Various types of equipment are used to furnish measurement data.

Orifice measurement station. *Orifice meter station* designates a complete measurement installation consisting of meter tube, orifice plate, pressure taps, and a differential pressure gauge (orifice meter). In the measurement of gas by orifice meter, the chart contains records of the differential and the static pressure. From these records, the quantity of gas measured is determined by the use of the formula in AGA Report No. 3.

Turbine meter. The name *turbine meter* is given to those meters in which there is a turbine wheel or rotor made to rotate by the flowing fluid—gas in the present case. The gas turbine meter is a velocity device which measures the volume of gas flowing through it parallel to the rotor action of the meter and by inferring that the speed of rotation of the rotor is directly proportional to the rate of flow. Measurement of fuel gas by turbine meters is covered in AGA Report No. 7.

Diaphragm displacement meter. This meter differs from the orifice type in that the gas actually flows through the meter and fills a definite measuring unit. Diaphragm meters have a maximum capacity ranging from approximately 75 to 11,000 ft^3/h. Industrial metering is generally performed at elevated pressure, and a pressure factor is applied to correct the metered volume to standard cubic feet. Correction for the elevated pressure can be made with a volume and pressure chart, an integrating device, a pressure-compensated index, or a fixed factor.

Volume and pressure gauge. The chart on the volume and pressure gauge is rotated either by a clock or the movement of the meter. The pressure pen records the gauge pressure of the gas in the meter and a volume pen indicates the quantity of the metered volume before pressure correction.

Temperature recorder. The temperature recorder measures and records on a chart the flowing gas temperature, usually in degrees Fahrenheit or percent of temperature chart range. Many times a temperature pen is installed as the third pen on the orifice meter to provide a temperature record along with differential and static pressure record.

Gravitometer. Recording gravitometers are used to obtain specific gravities when a continuous record is desired. It is a direct-weighing-type instrument and is constructed to measure the difference in the weight of a column of gas and an equal column of dry air. This value is transmitted through a linkage to the chart and recorded as the specific gravity of the gas continuously passing through the instrument.

Recording calorimeter.   A calorimeter is one of the instruments that is used to continuously record the heating value of gas. The calorimeter consists of two major parts—the tank unit in which a small continuous sample of gas is measured and burned and the recorder unit. The heat value is recorded as Btu, dry or saturated, per cubic foot for a specified pressure base.

Gas sampler.   Continuous gas samplers are used to obtain samples of gas continuously for a period up to a month. This method is used when it is not economically feasible to have on-line calorimeters or chromatographs at each metering location. The sample of gas in the collection container is taken to the laboratory for gas analysis.

Gas sample container.   Containers of stainless steel are usually used to obtain a sample of gas from a single source or to check gas from various sources by spot sampling or by continuous sampling over a period. A spot sample represents the gas stream at the instant of sampling. However, if gas composition does not vary, it can be assumed to be representative of the gas stream in general.

Gas chromatography equipment.   The chromatograph is a widely accepted analytical test instrument that is used to determine individual components in a gas stream. From this analysis of the tested gas, the gross heating value is calculated. The ideal values are adjusted to actual conditions by applying a compressibility factor. A chromatographic analysis can also provide the specific gravity (relative density) of the gas as well as the amount of liquefiable hydrocarbon available in the gas stream.

Electronic instruments.   These are devices that convert physical variables such as differential pressure, static pressure, temperature, specific gravity, and heating value to electrical signals. These signals are then processed to display the variable, to calculate flow, or control certain physical variables with sophisticated electronic devices.

Electronic flow computer.   This is an electronic process computer that obtains signals from electronic instruments and processes them to calculate flow rate (Mscf) or energy rate (MMBtu) based on contractual standards (for example, AGA-3, AGA-7, NX-19). In addition, some flow computers can control flow or pressure, provide regular printouts, store certain amounts of historical information, and communicate with a remote host computer.

Electronic gas measurement (EGM) system. This is the total system, consisting of electronic instruments, field flow computers, host computer, and communication network (microwave, radio, leased telephone, satellite links). Such a system can provide for remote monitoring and/or control, automated billing for custody transfer volume, online system analysis, etc. Although capabilities may vary, basic concepts remain the same for such systems.

1.3 Measurement Office Equipment

General

Certain kinds of office equipment are used to process measurement data generated in the field from physical variables. With the ever-increasing cost of natural gas, more emphasis is being placed on the speed and accuracy of gas measurement systems.

Chart integrator. This machine is manually controlled by the operator, who manipulates two control arms so that the two pen arms trace simultaneously the instantaneous values as the orifice meter chart rotates on the chart table, which is motor-driven by a foot-controlled feed. From these values, the chart extension, percent of pressure, and flow time are calculated and printed on paper tape and on each chart. Circuitry is also available to enter the data directly to the computer for volume calculation.

Chart scanner. This is an optical scanning device designed to integrate American, Foxboro, or Rockwell orifice meter charts. Each scanner is a complete system containing all optical, electronic, and mechanical components necessary to integrate, display, and print the chart's extension, pressure, and flow time. Advances in technology have resulted in modern optical scanners that process orifice as well as positive displacement meter charts and take more samples than older scanners, thereby improving accuracy of calculation. These machines are interfaced with computers.

Circular chart averager. This unit is used to obtain the average flowing temperature from the temperature charts.

Strip chart averager. This unit is used to obtain the average heat content (in Btu) from the calorimeter strip charts.

Computer data entry terminal. This unit is used to enter into the measurement data base the following data:

1. Station master

2. Orifice meter master

3. Displacement meter master

4. Gravity charts

5. Btu charts

6. Orifice meter reports

7. Positive meter reports

8. Orifice charts

9. Positive charts

10. Gas analysis reports

11. Miscellaneous entries

1.4   Processing of Measurement Data

The chart processing center serves as the central collection point for all reports and data that relate to the actual measurement of natural gas.

It is the responsibility of the chart processing group to accurately assemble and record all data concerning the primary, secondary, and auxiliary elements and equipment and to combine these factors with the integrated and averaged values of the chart recording to determine the volume and heat value of gas purchased, exchanged, stored, sold, transferred from one system to another, and used in company operation.

Flow chart of measurement data

The attached block diagram shows flow of measurement data (Fig. 1.1).

Station master

The station master record is set up from the gas contract and the meter installation report.

Meter master

The meter master record is set up from the meter installation report. The master record is the key to volume determination, and its preparation is normally checked very carefully.

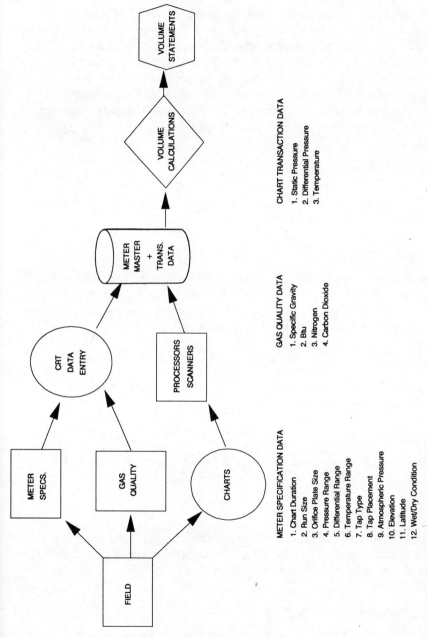

Figure 1.1 Flow chart of measurement data.

METER SPECIFICATION DATA

1. Chart Duration
2. Run Size
3. Orifice Plate Size
4. Pressure Range
5. Differential Range
6. Temperature Range
7. Tap Type
8. Tap Placement
9. Atmospheric Pressure
10. Elevation
11. Latitude
12. Wet/Dry Condition

GAS QUALITY DATA

1. Specific Gravity
2. Btu
3. Nitrogen
4. Carbon Dioxide

CHART TRANSACTION DATA

1. Static Pressure
2. Differential Pressure
3. Temperature

Chart censoring

Charts are censored for any irregularities. There are numerous discrepancies the censor must check for: time, differential pressure, flowing pressure, chart zero, plate size, meter station number, and general chart condition. The censoring clerk is a person who is usually familiar with the measuring station and can pick up discrepancies with a minimal amount of referral work. When an error in the recording or data is suspected, the chart censor will contact field personnel to obtain or confirm data. After the orifice chart is thoroughly censored, it is passed on to the chart cleanup personnel.

Chart cleanup

The chart is prepared for the chart scanner or it is scheduled for the integrator. Frequently touch-up is required on the differential pressure and static pressure recordings.

Integrator operator

The measurement clerk who manually integrates the orifice meter charts is the key to accurate volume determination. After the charts are integrated, the clerk will prepare a batch control card. The charts are then forwarded to computer data entry.

Data entry clerk

The data entry clerk uses various data entry programs to input measurement data.

Verification clerk

Error and warning messages from data processing are checked by the verification clerk.

File clerk

After error and warning messages are checked out, the charts are filed by station number. At the end of the month, some of the charts are pulled and mailed to other companies for auditing.

1.5 Orifice Flow Calculation

In the United States, AGA Report No. 3 is used as a standard for orifice metering. Equations prescribed in the report are used in various forms to calculate flow. Although the International Standards Organization (ISO) uses a different equation, the results are not significantly different. At the same time there are various assumptions and uncertainties involved in both sets of equations. Presently the AGA, American Petroleum Institute (API), American National Standards Institute (ANSI), and ISO are working together to arrive at an acceptable equation for all. The new set of equations, as approved by AGA, GPA, API, and ANSI recently, fits the experimental data better than existing equations. However, as of this time the most commonly used equations are as obtained from AGA Report No. 3, second edition, September 1985.

Method 1: AGA Report No. 3, second edition, September 1985

In the measurement of most gases, and especially natural gas, the general practice is to express the flow in cubic feet per hour at some specified reference base condition of pressure and temperature. For computational convenience the flow equation is written as

$$Q_v = C'(h_w P_f)^{0.5} \qquad (1.2)$$

and

$$C' = F_b\, F_r\, Y F_{pb}\, F_{tb}\, F_{tf}\, F_{gr}\, F_{pv} \qquad (1.3)$$

where Q_v = volume flow rate, ft³/h, at base condition
  h_w = differential pressure, inH$_2$O at 60°F
  P_f = absolute static pressure, lb/in²; subscript 1 (P_{f1}) is used when static pressure is measured at the upstream tap, subscript 2 (P_{f2}) is used when static pressure is measured at the downstream tap
  C' = orifice flow coefficient
  F_b = basic orifice factor
  F_r = Reynolds number factor
  Y = expansion factor
  F_{pb} = pressure base factor
  F_{tb} = temperature base factor
  F_{tf} = flowing temperature factor
  F_{gr} = real-gas relative density factor
  F_{pv} = supercompressibility factor

There are a few other factors such as manometer factor F_m, location factor F_l, and orifice thermal expansion factor F_a. Their influence on orifice calculation is not significant and, for all practical purposes, they can be assumed to be 1. F_m and F_l apply only with mercury manometer-type gauges.

The various factors to calculate C' can be calculated by the following equations.

Basic orifice factor

$$F_b = 338.178d^2K_o \tag{1.4}$$

where K_o is the flow coefficient when the orifice Reynolds number is infinitively large:

$$K_o = \frac{K_e}{1 + \dfrac{15E}{d(10^6)}} \tag{1.5}$$

and

$$E = d(830 - 5000\beta + 9000\beta^2 - 4200\beta^3 + B) \tag{1.6}$$

$$B = \frac{530}{D^{0.5}} \quad \text{for flange tap} \tag{1.7}$$

$$B = \frac{875}{D} + 75 \quad \text{for pipe tap} \tag{1.8}$$

For a flange tap,

$$K_e = 0.5993 + \frac{0.007}{D} + \left(0.364 + \frac{0.076}{D^{0.5}}\right)\beta^4 + 0.4\left(1.6 - \frac{1}{D}\right)^5$$

$$\left[\left(0.07 + \frac{0.5}{D}\right) - \beta\right]^{5/2} - \left(0.009 + \frac{0.034}{D}\right)(0.5 - \beta)^{3/2}$$

$$+ \left[\frac{65}{D^2} + 3\right][\beta - 0.7]^{5/2} \tag{1.9}$$

For a pipe tap,

$$K_e = 0.5925 + \frac{0.0182}{D} + \left(0.440 - \frac{0.06}{D}\right)\beta^2 + \left(0.935 + \frac{0.225}{D}\right)\beta^5$$

$$+ 1.35\beta^{14} + \frac{1.43}{D^{0.5}}(0.25 - \beta)^{5/2} \tag{1.10}$$

where D = meter tube diameter, in
  d = orifice diameter, in
  $\beta = d/D$
  K_e = flow coefficient when orifice Reynolds number R_d is equal to $d(10^6)/15$

Reynolds number factor

$$F_r = 1 + \frac{E}{R_d} \tag{1.11}$$

$$R_d = \frac{V_f d}{12\mu}\, \gamma \tag{1.12}$$

where μ = viscosity for natural gas, 0.0000069 lb/ft.

$$V_f = 11.125\, K\, \sqrt{\frac{h_w T_f}{P_f G}} \tag{1.13}$$

$$\gamma = 2.702\, \frac{P_f G}{T_f}\, (F_{pv})^2 \tag{1.14}$$

Combining Eqs. (1.13) and (1.14) in Eq. (1.12) gives

$$R_d = (3.6304 \times 10^5)dK\frac{h_w P_f G}{T_f}(F_{pv})^2 \tag{1.15}$$

where

$$K = K_o\left(1 + \frac{E}{R_d}\right) \tag{1.16}$$

K and R_d can be determined by iterative calculation. However, an average value of K, or of K_o in place of K, can be used in Eq. (1.15) with no significant inaccuracy.

Expansion factor

Using upstream static pressure, we write the following equations for expansion factor Y.

For flange taps,

$$Y_1 = 1 - (0.41 + 0.354\beta^4)\frac{x_1}{k} \tag{1.17}$$

For pipe taps,

$$Y_1 = 1 - [0.333 + 1.145(\beta^2 + 0.7\beta^5 + 12\beta^{13})]\frac{x_1}{k} \tag{1.18}$$

and

$$x_1 = \frac{P_{f1} - P_{f2}}{P_{f1}} = \frac{h_w}{27.707 P_{f1}} \tag{1.19}$$

Using downstream static pressure, we write the following equations.

For flange taps,

$$Y_2 = Y_1 \left(\frac{1}{1 - x_1} \right)^{0.5} \tag{1.20}$$

$$= (1 + x_2)^{0.5} - (0.41 + 0.35\beta^4) \frac{x_2}{k(1 + x_2)^{0.5}} \tag{1.21}$$

For pipe taps,

$$Y_2 = (1 + x_2)^{0.5}$$

$$- [0.333 + 1.145(\beta^2 + 0.7\beta^5 + 12\beta^{13})] \frac{x_2}{k(1 + x_2)^{0.5}} \tag{1.22}$$

and

$$x_2 = \frac{P_{f1} - P_{f2}}{P_{f2}} = \frac{h_w}{27.707 P_{f2}} \tag{1.23}$$

where Y_1 = expansion factor based on static pressure measured at the upstream tap

Y_2 = expansion factor based on static pressure measured at the downstream tap

h_w = differential pressure, inH_2O at 60°F

P_{f1} = static pressure at upstream pressure tap, lb/in^2 absolute

P_{f2} = static pressure at downstream pressure tap, lb/in^2 absolute

k = C_p/C_v, the ratio of the specific heats of the gas at constant pressure and constant volume at flowing conditions (a value of 1.3 is commonly used)

Pressure base factor

$$F_{pb} = 14.73/P_b \tag{1.24}$$

where P_b = required (contract) base pressure, lb/in^2 absolute.

Temperature base factor

$$F_{tb} = T_b/519.67 \tag{1.25}$$

where T_b = required (contract) base temperature, °R.

Flowing temperature factor

$$F_{tf} = \left(\frac{519.67}{T_f}\right)^{0.5}$$ (1.26)

where T_f = actual flowing temperature of gas, °R.

Real-gas relative density (specific gravity) factor

$$F_{gr} = \left(\frac{1}{G_r}\right)^{0.5}$$ (1.27)

where G_r = real-gas relative density.

Supercompressibility factor

For orifice meter measurement of gases, the effect of compressibility equates to the factor $(1/Z)^{0.5}$. This has been termed the *supercompressibility* F_{pv} of the gas.

Method 2: AGA Report No. 3, Part 1, third edition, October, 1990

The practical orifice flow equation used in this standard is:

$$q_m = N_1 C_d E_v Y d^2 \sqrt{\rho_{t,p}\Delta P}$$

The volumetric flow rate at standard (base) conditions is given by:

$$Q_v = q_m/\rho_b$$

where C_d = orifice plate coefficient of discharge
 d = orifice plate bore diameter calculated at flowing temperature (T_f)
 ΔP = orifice differential pressure (inH$_2$O in U.S. units)
 E_v = velocity of approach factor
 N_1 = unit conversion factor = 0.0997424 in U.S. units
 q_m = mass flow rate
 $\rho_{t,p}$ = density of the fluid at flowing condition (P_f,T_f)
 ρ_b = density of the fluid at base conditions
 Y = expansion factor
 Q_v = volume flow rate at base condition

Coefficient of discharge (C_d) equation. This equation was developed by Reader-Harris/Galagher (RG) for concentric, square, edged, flange-tapped orifice meter and is shown below.

$$C_d(\text{FT}) = C_i(\text{FT}) + 0.000511 \left[\frac{10^6\beta}{\text{Re}_D}\right]^{0.7} + (0.0210 + 0.0049A)\beta^4 C$$

$$C_i(\text{FT}) = C_i(\text{CT}) + \text{TapTerm}$$

$$C_i(\text{CT}) = 0.5961 + 0.0291\beta^2 - 0.2290\beta^8 + 0.003(1 - \beta)M_1$$

$$\text{Tap Term} = \text{Upstrm} + \text{Dnstrm}$$

$$\text{Upstrm} = [0.0433 + 0.0712e^{-8.5L_1} - 0.1145e^{-6.0L_1}](1 - 0.23A)B$$

$$\text{Dnstrm} = -0.0116[M_2 - 0.52M_2^{1.3}]\beta^{1.1}(1 - 0.14A)$$

Also,

$$B = \frac{\beta^4}{1 - \beta^4}$$

$$M_1 = \max\left(2.8 - \frac{D}{N_4}, 0.0\right)$$

$$M_2 = \frac{2L_2}{1 - \beta}$$

$$A = \left[\frac{19,000\beta}{\text{Re}_D}\right]^{0.8}$$

$$C = \left[\frac{10^6}{\text{Re}_D}\right]^{0.35}$$

where β = diameter ratio = d/D
$C_d(\text{FT})$ = coefficient of discharge at a specified pipe Reynolds number for flange-tapped orifice meter
$C_i(\text{FT})$ = coefficient of discharge at infinite pipe Reynolds number for flange-tapped orifice meter
$C_i(\text{CT})$ = coefficient of discharge at infinite pipe Reynolds number for corner-tapped orifice meter
d = orifice plate bore diameter calculated at T_f
D = meter tube internal diameter calculated at T_f
e = Napierian constant = 2.71828
L_1 = dimensionless correction for the tap location = $L_2 = N_4/D$ for flange taps
N_4 = 1.0 when D is in inches
     = 25.4 when D is in millimeters
Re_D = pipe Reynolds number

Reynolds number (Re_D). The RG equation uses the pipe Reynolds number. The pipe Reynolds number can be calculated using the following equation.

$$\text{Re}_D = \frac{4q_m}{\pi \mu D}$$

The pipe Reynolds number equation used in this standard is in a simplified form that combines the numerical constants and unit conversion constants:

$$\text{Re}_D = \frac{N_2 q_m}{\mu D}$$

where D = meter tube internal diameter calculated at flowing temperature (T_f)

     μ = absolute viscosity of fluid (use poise in U.S. units)

     N_2 = unit conversion factor = 227.375 in U.S. units

     π = universal constant = 3.14159

     q_m = mass flow rate

     Re_D = pipe Reynolds number

Velocity of approach factor (E_v). The velocity of approach factor, E_v, is calculated as follows:

$$E_v = \frac{1}{\sqrt{1 - \beta^4}}$$

and

$$\beta = d/D$$

where d = orifice plate bore diameter calculated at flowing temperature (T_f)

     D = meter tube internal diameter calculated at flowing temperature (T_f)

Orifice plate bore diameter (d). The orifice plate bore diameter, d, is defined as the diameter at flowing conditions and can be calculated using the following equation:

$$d = d_r[1 + \alpha_1(T_f - T_r)]$$

where α_1 = linear coefficient of thermal expansion for the orifice plate material

     d = orifice plate bore diameter calculated at flowing temperature (T_f)

d_r = reference orifice plate bore diameter at T_r

T_f = temperature of the fluid at flowing conditions

T_r = reference temperature of the orifice plate bore diameter

Note: α, T_f, and T_r must be in consistent units. For the purpose of this standard, T_r is assumed to be 68°F (20°C).

Meter tube internal diameter (D).   The meter tube internal diameter, D, is defined as the diameter at flowing conditions and can be calculated using the following equation:

$$D = D_r[1 + \alpha_2(T_f - T_r)]$$

where α_2 = linear coefficient of thermal expansion for the meter tube material

D = meter tube internal diameter calculated at flowing temperature (T_f)

D_r = reference meter tube internal diameter at T_r

T_f = temperature of the fluid at flowing conditions

T_r = reference temperatute of the meter tube internal diameter

Note: α, T_f, and T_r must be in consistent units. For the purpose of this standard, T_r is assumed to be 68°F (20°C).

In U.S. units the linear coefficient of thermal expansion is as follows: 304 and 316 stainless steel—0.00000925; monel—0.00000795; and carbon steel—0.00000620.

Expansion factor (Y) for flange-tapped orifice meters.   The expansion factor, Y, is defined as follows:

$$Y = \frac{C_{d_1}}{C_{d_2}}$$

where C_{d_1} = coefficient of discharge from compressible fluids tests

C_{d_2} = coefficient of discharge from incompressible fluids tests

Within the limits of this standard's application, it is assumed that the temperatures of the fluid at the upstream and downstream differential sensing taps are identical for the expansion factor calculation.

The application of the expansion factor is valid as long as the following dimensionless pressure ratio criteria are followed:

$$0 < \frac{\Delta P}{N_3 P_{f1}} < 0.20$$

or

$$0.8 < \frac{P_{f2}}{P_{f1}} < 1.0$$

where ΔP = orifice differential pressure
N_3 = unit conversion factor = 1000.00 in U.S. units
P_f = absolute static pressure at the pressure tap
P_{f1} = absolute static pressure at the upstream pressure tap
P_{f2} = absolute static pressure at the downstream pressure tap

Although use of the upstream or downstream expansion factor equation is a matter of choice, the upstream expansion factor is recommended because of its simplicity. If the upstream expansion factor is chosen, then the determination of the flowing fluid compressibility should be based on the upstream absolute static pressure, P_{f1}. Likewise, if the downstream expansion factor is selected, then the determination of the flowing fluid compressibility should be based on the downstream absolute static pressure, P_{f2}.

The expansion factor equation for flange taps is applicable over a β range of 0.10–0.75.

Upstream expansion factor (Y_r). If the absolute static pressure is taken at the upstream differential pressure tap, then the value of the expansion factor, Y_1, shall be calculated as follows:

$$Y_1 = 1 - (0.41 + 0.35\beta^4)\frac{x_1}{k}$$

When the upstream static pressure is measured,

$$x_1 = \frac{\Delta P}{N_3 P_{f1}}$$

When the downstream static pressure is measured,

$$x_1 = \frac{\Delta P}{N_3 P_{f1} + \Delta P}$$

where ΔP = orifice differential pressure
k = isentropic exponent
N_3 = unit conversion factor (27.707 if ΔP, inH_2O)
P_{f1} = absolute static pressure at the upstream pressure tap
P_{f2} = absolute static pressure at the downstream pressure tap
x_1 = ratio of differential pressure to absolute static pressure at the upstream tap
x_1/k = upstream acoustic ratio
Y_1 = expansion factor based on the absolute static pressure measured at the upstream tap

Downstream expansion factor (Y_2). The value of the downstream expansion factor, Y_2 shall be calculated using the following equation:

$$Y_2 = Y_1 \sqrt{\frac{P_{f1}Z_{f2}}{P_{f2}Z_{f1}}}$$

or

$$Y_2 = \left[1 - (0.41 + 0.35\beta^4)\frac{x_1}{k}\right] \sqrt{\frac{P_{f1f2}}{P_{f2}Z_{f1}}}$$

When the upstream static pressure is measured,

$$x_1 = \frac{\Delta P}{N_3 P_{f1}}$$

When the downstream static pressure is measured,

$$x_1 = \frac{\Delta P}{N_3 P_{f1} + \Delta P}$$

where ΔP = orifice differential pressure
k = isentropic exponent
N_3 = unit conversion factor
P_{f1} = absolute static pressure at the upstream pressure tap
P_{f2} = absolute static pressure at the downstream pressure tap
x_1 = ratio of differential pressure to absolute static pressure at the upstream tap
x_1/k = upstream acoustic ratio
Y_1 = expansion factor based on the absolute static pressure measured at the upstream tap
Y_2 = expansion factor based on the absolute static pressure measured at the downstream tap
Z_{f1} = fluid compressibility at the upstream pressure tap
Z_{f2} = fluid compressibility at the downstream pressure tap

1.6 Supercompressibility Calculation

At this time, supercompressibility factor F_{pv} is most commonly determined by using the formulas given in the AGA manual for the determination of supercompressibility factors for natural gas, Pipeline Research Committee (PAR) research project NX-19. F_{pv} is also dealt with in great detail in AGA Report No. 8 with improved methods of calculation.

Method 1: NX-19

$$F_{pv} = \frac{\sqrt{\dfrac{B}{D}} - D + \dfrac{n}{3\pi}}{1 + \dfrac{0.00132}{\tau^{3.25}}} \qquad (1.28)$$

where

$$B = \frac{3 - mn^2}{9m\pi^2} \qquad (1.29)$$

$$m = 0.0330378\tau^{-2} - 0.0221323\tau^{-3} + 0.0161353\tau^{-5} \qquad (1.30)$$

$$n = \frac{0.265827\tau^{-2} + 0.0457697\tau^{-4} - 0.133185\tau^{-1}}{m}$$

$$\pi = \frac{P_{adj} + 14.7}{1000} \qquad (1.31)$$

$$\tau = \frac{T_{adj} + 460}{500} \qquad (1.32)$$

$$P_{adj} = \frac{156.47P_{f1}}{160.8 - 7.22G_r + (M_c - 0.392M_n)} \qquad (1.33)$$

$$T_{adj} = \frac{226.29T_f}{99.15 + 211.9G_r - (M_c + 1.681M_n)} - 460 \qquad (1.34)$$

$$D = (b + \sqrt{b^2 + B^3})^{1/3} \qquad (1.35)$$

$$b = \frac{9n - 2mn^3}{54m\pi^3} - \frac{E}{2m\pi^2} \qquad (1.36)$$

where G_r = specific gravity of flowing gas
$\quad M_c$ = mole percent carbon dioxide
$\quad M_n$ = mole percent nitrogen
$\quad P_{f1}$ = static pressure, lb/in^2 gauge
$\quad T_f$ = flowing temperature, °R
$\quad P_{adj}$ = adjusted pressure for F_{pv} equation, lb/in^2 gauge
$\quad T_{adj}$ = adjusted temperature for F_{pv} equation, °F

F_{pv} can be also determined (and interpolated) from AGA NX-19 tables, which are given for a range of adjusted temperature and pressure for 0.6 specific gravity hydrocarbon gas.

E can be calculated by the following formulas according to the range of applicability as shown in Fig. 1.2.

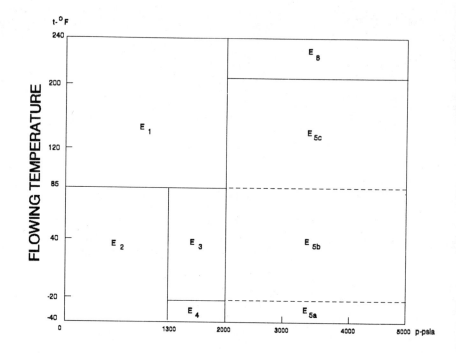

PRESSURE

Figure 1.2   Range of applicability of the parameter E.

Equations for parameter E

$$E_1 = 1 - 0.00075\pi^{2.3}e^{-20(\tau - 1.09)} - 0.0011(\tau - 1.09)^{0.5}\pi^2$$

$$\times [2.17 + 1.4\,(\tau - 1.09)^{0.5} - \pi]^2 \quad (1.37)$$

Pressure range: π, 0 to 2; P, 0 to 2000 lb/in^2 absolute.
Temperature range: τ, 1.09 to 1.4; T, 85°F to 240°F.

$$E_2 = 1 - 0.00075\pi^{2.3}(2 - e^{-20(1.09 - \tau)})$$

$$- 1.317(1.09 - \tau)^4(\pi)(1.69 - \pi^2) \quad (1.38)$$

Pressure range: π, 0 to 1.3; P, 0 to 1300 lb/in^2 absolute.
Temperature range: τ, 0.84 to 1.09; T, −40°F to 85°F.

$$E_3 = 1 - 0.00075(\pi)^{2.3}(2 - e^{-20(1.09 - \tau)}) + 0.455$$

$$\times [200(1.09 - \tau)^6 - \cdots - 0.03249(1.09 - \tau)$$

$$+ 2.0167(1.09 - \tau)^2 - 18.028\,(1.09 - \tau)^3 + \cdots$$

$$+ 42.844(1.09 - \tau)^4](\pi - 1.3)\,[1.69(2)^{1.25} - \pi^2] \quad (1.39)$$

Pressure range: π, 1.3 to 2.0; P, 1300 to 2000 lb/in^2 absolute.
Temperature range: τ, 0.88 to 1.09; T, $-20°F$ to 85°F.

E_4 = same as E_3 except exponent 1.25 is changed to

$$1.25 + 80\,(0.88 - \tau)^2 \quad (1.40)$$

Pressure range: π, 1.3 to 2.0; P, 1300 to 2000 lb/in^2 absolute.
Temperature range: τ, 0.84 to 0.88; T, $-40°F$ to $-20°F$.

$$E_{5a} = E_4 - Y \quad (1.41)$$

Pressure range: π, 2.0 to 5.0; P, 2000 to 5000 lb/in^2 absolute.
Temperature range: τ, 0.84 to 0.88; T, $-40°F$ to $-20°F$.

$$E_{5b} = E_3 - Y \quad (1.42)$$

Pressure range: π, 2.0 to 5.0; P, 2000 to 5000 lb/in^2 absolute.
Temperature range: τ, 0.88 to 1.09; T, $-20°F$ to 85°F.

$$E_{5c} = E_1 - Y \quad (1.43)$$

Pressure range: π, 2.0 to 5.0; P, 2000 to 5000 lb/in^2 absolute.
Temperature range: τ, 1.09 to 1.32; T, 85°F to 200°F.

$$E_6 = E_{5c} - U \quad (1.44)$$

Pressure range: π, 2.0 to 5.0; P, 2000 to 5000 lb/in^2 absolute.
Temperature range: τ, 1.32 to 1.40; T, 200°F to 240°F.

$$Y = A(\pi - 2) + A_1(\pi - 2)^2 + A_2(\pi - 2)^3 + A_3(\pi - 2)^4 \quad (1.45)$$

where

$$A = 1.71720 - 2.33123\tau - 1.56796\tau^2 + 3.47644\tau^3 - 1.28603\tau^4$$

$$A_1 = 0.016299 - 0.028094\tau + 0.48782\tau^2 - 0.728221\tau^3 + 0.27839\tau^4$$

$$A_2 = -0.35978 + 0.51419\tau + 0.16453\tau^2 - 0.52216\tau^3 + 0.19687\tau^4$$

$$A_3 = 0.075255 - 0.10573\tau - 0.058598\tau^2 + 0.14416\tau^3 - 0.054533\tau^4$$

$$U = (\tau - 1.32)^2(\pi - 2)[3 - 1.483(\pi - 2)$$
$$- 0.10\,(\pi - 2)^2 + 0.0833(\pi - 2)^3] \quad (1.46)$$

Note: π and τ are defined by Eqs. (1.31) and (1.32), respectively. The napierian constant $e \approx 2.7182818$.

**Method 2: Supercompressibility
determination using method prescribed in
AGA Report No. 8**

This method is required to be used if the latest AGA-3 method is used
for orifice flow calculation.

Supercompressibility is given by the relation:

$$F_{pv} = \frac{Z_b}{Z}$$

where F_{pv} = supercompressibility factor
Z = compressibility factor at conditions of interest
Z_b = compressibility factor at base conditions

The compressibility factor Z is defined by:

$$Z = \frac{P}{dRT}$$

The equation for absolute P is given by:

$$P = RTd + BRTd^2 + CRTd^3 + DRT^4 + ERTd^6$$
$$+ A_1 RTd^3 (1 + A_2 d^2) \exp(-A_2 d^2)$$

where Z = compressibility factor of gas
P = absolute pressure
R = gas constant
T = absolute temperature
d = molar density of gas
B, C, D, E, A_1, A_2 = coefficients which are functions of temperature
and composition

When the temperature, pressure, and composition of a gas are
known, the coefficients, B, C, D, E, A_1, and A_2 can be determined and
the above equation solved for the molar density d.

Methods for determining B, C, D, E, A_1, and A_2 and for solving for
molar density are given in detail in AGA Report No. 8, December 15,
1985 edition. Readers should refer to that report.

1.7   Flow Calculation for Turbine or Positive Displacement Meter

A turbine meter is a velocity-measuring device, and, as such, rotor
speed is proportional to the flow rate. Since rotor revolutions are
counted to give volume at flowing condition, the count has to be corrected for specified base conditions. In the same manner, a positive

displacement meter displaces a known volume at flowing conditions, and this volume has to be corrected for specified base conditions. The equations for calculating volumetric flow rate for a turbine or positive displacement meter are as follows:

$$V_b = V_f\left(\frac{P_f}{P_b}\right)\left(\frac{T_b}{T_f}\right)(F_{pv})^2 \tag{1.47}$$

$$Q_b = Q_f\left(\frac{P_f}{P_b}\right)\left(\frac{T_b}{T_f}\right)(F_{pv})^2 \tag{1.48}$$

where Q_b = flow rate at base conditions = V_b/t
Q_f = flow rate at flowing conditions = V_f/t
V_b = volume at base conditions
V_f = volume at flowing conditions = counter difference on mechanical output = total pulses × $1/K$ on electrical output
t = time
K = pulses per cubic foot

1.8   Other Flow Meter Calculations

Annubar

The Annubar is another inferential device that utilizes a form of the classical Bernoulli energy balance equation to determine flow rate. The differential pressure produced by the Annubar is consistent and uniform for a given pipe size and flow condition. Therefore, only the operating range of the instrument or control system needs to be sized. However, for the same pipe size, pressure, and flow rate, an orifice meter gives a much higher differential. Therefore, Annubar measurement has lower resolution. There is no established industry standard, and at present it is not generally accepted for custody transfer measurement. However, because of lower cost and operating ease, it is used for check measurement, certain control schemes, and gas odorization purposes.

Volumetric flow rate of gas at standard conditions is given by

$$Q_n = 7.897 SND^2 F_a V_a \frac{\sqrt{\gamma_f}}{\gamma_b}\sqrt{h_n} \tag{1.49}$$

where Q_n = flow rate in desired units
N = constant including $2g$ (gravity acceleration), $\pi/4$, and conversion constant ($N = 45.46$ for Q_n in ft^3/h and h_n in inH$_2$O)
S = discharge coefficient for Annubar flow sensor (available from manufacturer)

D = inside diameter of pipe, in

F_a = expansion factor; ratio of inside area of pipe at flowing temperature to that at 68°F (F_a = 1 between +31°F and +106°F in steel pipe)

V_a = gas adiabatic compression factor (velocities below 12,000 ft/min use 1.00); equal to

$$\sqrt{\frac{k}{(k-1)}\frac{P_s}{(P_t-P_s)}\left(\left[\frac{P_t}{P_s}\right]^{(k-1)/k}-1\right)}$$

k = ratio of specific heats (k = 1.3 for natural gas)

P_s = upstream static pressure in consistent absolute units

P_t = total or impact pressure at element in consistent absolute units

γ_f = specific weight of flowing conditions, lb/ft³, including compressibility; equal to

$$\frac{P_f}{14.73} \times \frac{520}{460 + T_f} \times \gamma_b$$

P_f = flowing pressure = atmospheric pressure + line pressure, lb/in² gauge

T_f = flowing temperature, °F

γ_b = specific weight at base conditions, lb/ft³; equal to specific gravity of gas at base conditions times the weight of air (lb/ft³) at base conditions (air = 0.0765 lb/ft³ at 60°F, 14.73 lb/in² absolute)

h_n = differential pressure output of Annubar element in appropriate units

Volumetric flow rate using an Annubar can be also calculated using the familiar orifice-flow-type formula,

$$Q_s = C' \sqrt{\frac{hP_f}{T_f}} \tag{1.50}$$

where C' = $F_{NA}KD^2F_rF_{AA}Y_AF_{pb}F_{tb}F_GF_{pv}$

Q_s = flow rate, standard ft³/day

F_{NA} = conversion factor

K = Annubar flow coefficient

D = inside pipe diameter, in

F_r = Reynolds number factor

F_{AA} = pipe thermal expansion factor

Y_A = gas expansion factor

F_{pb} = pressure base factor
F_{tb} = temperature base factor
F_{gr} = specific gravity factor = $\sqrt{1/G_r}$
F_{pv} = supercompressibility factor
  h = differential pressure, inH_2O
P_f = flowing pressure, lb/in^2 absolute
T_f = flowing temperature, °R

Vortex-shedding flow meter

The vortex-shedding flow meter is another device that uses the well-known phenomenon of vortex shedding by a bluff body to determine fluid velocity and thereby infer flow rate. Fluid flowing through the flow meter body passes a specially shaped element which causes vortices to form and shed (separate) from alternate sides of the element at a rate proportional to the velocity of the fluid. These shedding vortices create an alternating differential pressure. This can also cause transverse oscillation of the element at the same frequency as the vortex shedding. By sensing the alternating differential pressure of the vortices (possibly by an acoustic method) or by sensing transverse oscillation of the element, one can determine vortex-shedding frequency. This frequency is related to fluid velocity by the following relation:

$$V = \frac{fD}{N_s} \qquad (1.51)$$

where D = characteristic dimension of the element (for a cylindrical object, D is the diameter)
  N_s = Strouhal number, assumed to be constant in the Reynolds number range from 4×10^2 to 10^5
  f = vortex-shedding frequency or Strouhal frequency
  V = velocity of fluid

Knowing the fluid velocity, inside diameter of the pipe, flowing pressure, and flowing temperature, one can calculate a flow rate at standard conditions using a flowing formula similar to that for a turbine or positive displacement meter:

$$Q = AV\left(\frac{P_f}{P_b}\right)\left(\frac{T_b}{T_f}\right)(F_{pv})^2 \qquad (1.52)$$

where A = inner cross-sectional area of the pipe. The vortex-shedding meter is gaining popularity in the gas industry. It is presently not being used for custody transfer. However, it can be a good operational meter.

1.9 Pipe Flow Calculation

Weymouth formula

$$Q = 433.488\frac{T_o}{P_o}\sqrt{\frac{(P_1^2 - P_2^2)D^{16/3}}{GTL}} \qquad (1.53)$$

or

$$Q = \frac{C}{\sqrt{L}}\sqrt{P_1^2 - P_2^2} \qquad \text{to determine rate of flow} \qquad (1.54)$$

$$P_1 = \sqrt{P_2^2 + L\left(\frac{Q}{C}\right)^2} \qquad \text{to determine upstream pressure}$$

$$\qquad (1.55)$$

$$P_2 = \sqrt{P_1^2 - L\left(\frac{Q}{C}\right)^2} \qquad \text{to determine downstream pressure}$$

$$\qquad (1.56)$$

$$C = \frac{Q\sqrt{L}}{\sqrt{P_1^2 - P_2^2}} \qquad \text{to determine pipe size} \qquad (1.57)$$

$$L = \left(\frac{C}{Q}\right)^2(P_1^2 - P_2^2) \qquad \text{to determine length of line} \qquad (1.58)$$

where Q = quantity of gas flowing, ft^3/day
　　　L = length of line, miles
　　　P_1 = inlet pressure lb/in^2 absolute
　　　P_2 = terminal pressure, lb/in^2 absolute

Coefficient C is derived by substituting the following constants in Eq. (1.53):

T_o = base measurement temperature of 60°F = 520°R

P_o = base measurement pressure (as indicated) = 14.90 or

$$14.73 \text{ lb/in}^2 \text{ absolute}$$

G = specific gravity of gas (air = 1.00) = 0.60

T = mean flowing temperature of 60°F = 520°R

D = inside diameter of pipe, in

Thus, when L = 1 mi at P_o = 14.90 lb/in^2 absolute,

$$C = 433.488\left(\frac{520}{14.90}\right)\sqrt{\frac{D^{16/3}}{(0.60)(520)(1)}} = 856.480D^{8/3} \quad (1.59)$$

and, when $L = 1$ mi at $P_o = 14.73$ lb/in^2 absolute,

$$C = 433.488\left(\frac{520}{14.73}\right)\sqrt{\frac{D^{16/3}}{(0.60)(520)(1)}} = 866.364D^{8/3} \quad (1.60)$$

1.10 Gas Loss Calculation

General definitions

In general, there are three types of gas losses: (1) blowdown; (2) drain; and (3) purge. Such losses can be intentional or accidental.

1. *Blowdown loss:* Uncontrolled gas loss through pipeline puncture or blowout until ruptured section is isolated by closing block valves.

2. *Drain loss:* Loss due to emptying (draining) the gas in the section of ruptured line or any action on line after it has been isolated by closing block valves.

3. *Purge loss:* An intentional gas loss incurred in replacing and/or putting section of line back in service. This is done by flowing gas to move out the entrained air in the section.

The method or formula used for blowdown calculation also depends on hole size:

1. *Large hole:* Hole that is sufficiently large that *all* the gas in the pipeline leaves through it. If the hole is not large, only a partial amount of the pipeline's flow rate will discharge through the hole, and the rest will continue on through the pipeline. How large a hole is required so that *all* gas goes out the hole? One way of knowing is to determine the direction of gas flow in the downstream pipeline by checking pressure at two points downstream of the break (if possible).

2. *Small hole:* A hole is considered to be small if only a part of the gas is discharged at the hole and the rest moves normally down the pipeline. To determine if hole is "small," find out the direction of gas flow in the downstream pipeline by same method as for a large hole, noting direction of pressure drop. Note that a small hole has a flow capacity limit due to the "choked flow" condition.

Formulas for gas loss calculation

Blowdown

    Large hole

$$V = Qt = (0.0361) \frac{D^{8/3}}{\sqrt{L}} \sqrt{P_1^2 - P_2^2} \times t \qquad (1.61)$$

where V = volume of gas lost, Mscf
          Q = flow rate, Mscf/h
          t = blowdown time, h
          D = inside pipe diameter, in
          L = length of section, mi
          P_1 = average inlet pressure, lb/in^2, for the blowdown period, obtained from any available upstream pressure gauge
          P_2 = atmospheric pressure, 14.7 lb/in^2 absolute

    Small hole

$$V = (1.178)D^2Pt \qquad (1.62)$$

where V = volume of gas lost, Mscf
          D = average diameter of hole, in
          P = estimated average pipeline pressure just upstream of hole, lb/in^2 absolute
          t = blowdown time, h

Drain

$$V = (0.001955)D^2LP_{ave} \qquad (1.63)$$

where V = volume of gas lost, Mscf
          D = inside diameter of pipe, in
          L = length of pipe section, mi
        P_{ave} = average pressure in the pipe section, lb/in^2 absolute

For rupture with a large hole, $P_{ave} = P_m$, otherwise

$$P_{ave} = \frac{P_1 + P_2}{2} \qquad (1.64)$$

$$P_m = 0.667\left(P_1 + \frac{P_2^2}{P_1 + P_2}\right)$$

where P_1 = initial flowing upstream pressure in section to be drained, lb/in^2 absolute, before closure of block valves
       P_2 = initial flowing downstream pressure in section to be drained, lb/in^2 absolute, before closure of block valves

Purge. In order for these equations to be accurate, the flow rate at the blow-off must be at critical velocity. Therefore, in order to perform purge operations effectively and efficiently, a pressure gauge should be installed just upstream of the blow-off valve. With this gauge, the pipeline's downstream pressure can be monitored to ensure that critical velocity is maintained across the blow-off valve. The pressure necessary to accomplish this is any pressure greater than 14.7 lb/in^2 gauge; 18 lb/in^2 gauge or slightly higher is recommended. For 18 lb/in^2 gauge, the absolute pressure is 18 + 14.7 = 32.7 lb/in^2, which is slightly more than twice atmospheric.

For purging, the following rules should be observed to maximize the accuracy of gas loss calculations and provide a *complete* purge:

1. Use gauges and observe pressures. Note purge gas (injection) pressure P_1 and blow-off pressure P_2.

2. Begin counting purge time only after proper blow-off pressure has been set.

3. Monitor specific gravity at blow-off.

4. Continue purging even after calculated purge time is reached if specific gravity is changing.

5. Note total purge time if different from the calculated value. In order to obtain the best possible results from purge time equations, it should be assumed that P_2 will be approximately equal to 18 lb/in^2 gauge. P_1 can be calculated by rearranging the Weymouth formula and using the critical flow rate of the blow-off valve.

To calculate a purge time, first find flow rate through the blow-off valve by using the formula for critical velocity,

$$Q = KP_2 \qquad (1.65)$$

where Q = flow rate, Mscf/h
  K = flow coefficient, Mscf/(h · lb/in^2 absolute) (from Table 1.1)
  P_2 = pressure just upstream of blow-off valve, lb/in^2 absolute

Now, using the Q just obtained, use the rearranged Weymouth formula to find an estimated value of P_1 necessary to maintain this flow rate:

$$P_1^2 = \left(\frac{Q\sqrt{L}}{(0.0361)D^{8/3}} \right)^2 + P_2^2 \qquad (1.66)$$

P_1, P_2, and Q are now known, and the purge time can be calculated by

TABLE 1.1   One-Hour Blow-off Coefficients for Standard
Blow-off Sizes

Blow-off size, in	K, Mscf/(h · lb/in^2 absolute)
1	0.75
2	3.0
3	6.0
4	13.5
6	24.0
8	47.0
10	72.0

any of the following formulas. Equations (1.67) to (1.69) calculate
time. Equations (1.70) and (1.71) calculate volume.

The recommended purge time is $2T$. The minimum purge time in
minutes is

$$T = \frac{(0.117)D^2(L)P_m\sqrt{L}}{C\sqrt{P_1^2 - P_2^2}} \tag{1.67}$$

or

$$T = (0.078)D^2L \ \frac{1 + \dfrac{\dfrac{C_1^2}{K^2 + C_1^2}}{1 + \sqrt{\dfrac{C_1^2}{K^2 + C_1^2}}}}{C_1\sqrt{\dfrac{K^2}{K^2 + C_1^2}}} \tag{1.68}$$

where D = inside diameter of pipe, in
L = length of purge section, mi
$C_1 = C/\sqrt{L}$
$C = (0.0361)D^{8/3}$ (1-h Weymouth coefficient in Mscf/h · mi)
$P_m = 0.667[P_1 + P_2^2/(P_1 + P_2)]$, average pressure, lb/in^2 absolute
P_1 = pressure at upstream end of section, lb/in^2 absolute
P_2 = pressure at downstream end of section, lb/in^2 absolute (just
upstream of blow-off valve)
K = 1-h blow-off coefficient for standard blow-off sizes, Mscf/
(h · lb/in^2 absolute)

K coefficients have been empirically developed for plug valves. If K coefficients for ball valves are used, the estimated purge time will be reduced. However, in the interest of safety, it is recommended that the coefficients in Table 1.1 be used for all blow-off valves.

Then the purge time in minutes is

$$T = \frac{cV_1}{K}\left(\frac{P_{ave}}{P_2}\right) \tag{1.69}$$

where c = conversion constant = $60/14.73$ = 4.073

V_1 = actual volume of pipe section purged, thousand ft^3, where pipe section is assumed to be filled with air prior to purge

K = blow-off coefficient, Mscf/(h · lb/in^2 absolute)

P_{ave} = $(P_1 + P_2)/2$

P_1 = pressure at upstream end of section, lb/in^2 absolute

P_2 = pressure at downstream end of section, lb/in^2 absolute, just upstream of blow-off valve

The volume of gas lost, Mscf, is

$$V = \left(C_1 \sqrt{P_1^2 - \frac{C_1^2 P_1^2}{K^2 + C_1^2}} \times \frac{t}{60}\right) - V_1 \tag{1.70}$$

where V_1 = actual volume of pipe action purged, thousand ft^3, where pipe section is assumed to be filled with air prior to purge; equal to $(0.028798)D^2L$, D in inches, L in miles

P_1 = pressure at upstream end of section, lb/in^2 absolute

P_2 = pressure at downstream end of section, lb/in^2 absolute (just upstream of blow-off valve)

C_1 = $C/\sqrt{L}$, with C = $(0.0361)D^{8/3}$, D in inches, L in miles

K = 1-hour blow-off coefficient for standard blow-off sizes, Mscf/(h · lb/in^2 absolute)

t = Actual purge time, minutes

Another form is

$$V = \left(KP\frac{t}{60}\right) - V_1 \tag{1.71}$$

where V_1 = actual volume of pipe section purged, thousand ft^3, where pipe section is assumed to be filled with air prior to purge; equal to $(0.028798)D^2L$, D in inches, L in miles

P = pressure of downstream end of section, lb/in^2 absolute (just upstream of blow-off valve)

t = actual time of purge, minutes

K = 1-h blow-off coefficient for standard blow-off sizes, Mscf/(h · lb/in² absolute)

1.11 Gas Pipeline Blowdown Time

When it is necessary to take a natural gas line or segment of a line out of service for repair, it is important to know the time required for venting in order to estimate total down time. Tables 1.2 and 1.3 are used to estimate blowdown time.

The following Walworth Co. formula is also frequently used to calculate blowdown time in minutes:

$$T_m = 0.0588FP^{1/3}G^{1/2}(D/d)^2(L/n) \qquad (1.72)$$

where P = initial line pressure, lb/in² gauge

G = specific gravity of gas

D = inside diameter of line, in

d = inside diameter of blowdown stack, in

L = length of line, mi

n = number of stacks blowing simultaneously

F = choke factor that has the following values: ideal nozzle, 1.0; through port gate valve, 1.6; regular gate valve, 1.8; regular lubricated plug valve, 2.0; venturi lubricated plug valve, 3.2

TABLE 1.2   Blowdown time, Minutes, through Two Blow-Offs from 450 lb/in^2 Gauge

Line size, in	Blow-off size, in	Gate valves Section length, mi			Regular plug valves Section length, mi			Venturi plug valves Section length, mi		
		5	10	15	5	10	15	5	10	15
4	2	6	11	17	6	13	19			
6	2	13	26	39	14	29	43			
	3	6	12	18	7	13	20			
	4	3	7	10	4	8	11			
8	2	23	46	68	25	51	76			
	3	11	21	32	12	23	35			
	4	6	12	18	7	13	20			
10	2	35	70	—	39	78	—			
	3	16	32	49	18	36	54			
	4	9	18	27	10	21	31			
	6	4	9	13	5	10	16	6	13	19
12	3	24	48	73	27	53	80			
	4	14	27	41	15	31	46			
	6	6	13	19	8	16	23	10	19	29
14	3	28	57	85	31	62	94			
	4	16	32	48	18	36	54			
	6	8	15	23	9	18	27	11	23	34
	8	4	9	13	5	11	16	7	14	20
16	3	38	75	—	41	83	—			
	4	21	43	64	24	48	72			
	6	10	20	30	12	24	36	15	30	45
	8	6	12	17	7	14	21	9	18	27
18	4	27	54	82	31	61	92			
	6	13	26	38	15	31	46	19	38	58
	8	7	15	22	9	18	27	11	23	34
20	4	34	68	—	38	76	—			
	6	16	32	48	19	38	58	24	48	72
	8	9	19	28	11	22	34	14	29	43
22	4	41	82	—	46	93	—			
	6	19	39	58	23	47	70	29	58	87
	8	11	23	34	14	27	41	17	35	52
24	4	49	99	—	56	—	—			
	6	23	46	69	28	56	84	35	70	—
	8	14	27	41	16	33	49	21	42	63
26	6	27	55	82	33	66	99	41	82	—
	8	16	32	48	19	39	58	25	49	74
30	6	37	73	—	44	89	—	55	—	—
	8	21	43	64	26	52	78	33	66	99
34	8	27	55	82	33	66	99	42	85	—
36	8	31	62	92	37	75	—	48	95	—

For one blow-off, multiply approximate time from this table by 2.

For pressures other than 450 lb/in^2 gauge, multiply approximate time from this table by appropriate pressure multiplier from Table 1.3.

TABLE 1.3 Multipliers for Blowdown Time
for Pressures Other than 450 lb/in^2
Gauge

Pressure, lb/in^2 gauge	Multiplier
25	0.30
50	0.43
100	0.60
150	0.70
200	0.77
250	0.84
300	0.89
350	0.93
400	0.96
450	1.00
500	1.03
600	1.08
700	1.12
800	1.16
900	1.19
1000	1.22
1100	1.25
1200	1.27

2

Gas Measurement: Operational and Accounting Considerations

2.1 General

This chapter provides tables, nomograms, and procedures for day-to-day operational use. These are self-explanatory and easy to use. These will also help in measurement accounting by enabling one to estimate an error to make necessary adjustments to a volume statement. With the information given here, one can minimize errors and avoid possible failure by taking appropriate operational steps. The appropriate plate size can be selected for an orifice meter to provide more readable and accurate measurements. Possible run cycling can be avoided by selecting an appropriate setting for the run switching controller. Freeze-up condition can be avoided by proper heating or alcohol injection. Measurement errors can be estimated for various conditions. Gas loss can be estimated and accounted for by following proper procedures. In any case, this chapter is intended as a quick operational and measurement accounting guide.

In general, orifice meters, positive meters, and turbine meters are the most widely used meters. For reliable and accurate measurement, free liquid or liquid dropout must be avoided. Liquid buildup in front of the orifice plate will make orifice calculation inaccurate. The turbine meter is not intended for two-phase flow. The positive displacement meter will act as a separator and should not be used in a wet system, especially in a production facility. With a turbine meter, flow should be maintained at 10 percent of the meter capacity or above. When flow is below 3 percent of capacity, a turbine meter should be resized and replaced. Because of its inertial nature, a turbine meter is less accurate at lower flow rate. Although spin time is a good indicator of turbine meter performance, high-pressure testing of the turbine module at least every 2 years is recommended. An orifice meter tube in clean gas service may not require frequent inspection of the tube. However, it is recommended that an orifice meter tube be inspected every 3 years, especially at production facilities where dry, clean gas may not be flowing through the meter tube. Without a bypass or an additional run, it is not possible to drop a tube for inspection without shutting down. However, tools are available for making a preliminary inspection of a meter run in service using a probe with a light source.

In any case, every precaution must be taken to ensure accuracy and reliability of a meter. Pressure control, when used, should be upstream of the meter, for steady metering pressure. However, cost and meter run size may demand that pressure reduction and control be done downstream of the meter. Flow control may be used downstream of a meter. Necessary precautions must be taken to ensure that the instrument supply is adequate under extreme conditions. This is done by using proper heating, standby supply regulators, and an additional supply point.

It must be remembered that good operating practice, with good design engineering, makes a system work reliably.

From an operational standpoint, routine inspection/calibration of meter and regulator equipment must be done primarily for these reasons:

1. To assure reliable, accurate, and safe operation
2. To comply with contractual obligations
3. To comply with governmental requirements

Such inspection/calibration must be documented, reported, and archived. These are essential parts of auditing, dispute settlement, accounting adjustment, and governmental compliance. Overpressure protection reports and odorization reports are extremely important to meet state and federal requirements.

2.2 Orifice Metering

As a primary device, an orifice meter tube, as recommended by industry standards (AGA, ANSI, or API) with provision for inserting an orifice plate, remains the most widely used device for production and transmission measurement. This is expected to be the case for fluid measurement for years to come. Extensive research coupled with experimental verification, familiarity, and long-term practical applications have made this a device that cannot be equaled by any other for most high-pressure, large-volume applications. For distribution measurement, the positive displacement meter is the most widely used. With orifice metering, although further improvement of the formulas and better correlation with experimental data are continuing, the overall limiting factor appears to be that imposed by practical design and application, secondary devices, and nonideal conditions, including pulsations. Electronic replacements for secondary devices and continued research are reducing overall uncertainty.

In this chapter, Table 2.1 provides a method for sizing orifice runs quickly. Table 2.2 shows how to select an appropriate orifice plate for a desired differential reading. In general, the differential reading should be between 10 and 90 percent of the meter range. With a fixed orifice plate this provides 3:1 rangeability for a meter run. Rangeability can be improved by changing the plate, which requires human intervention. Rangeability can be also improved by installing multiple meter runs and automatic run switching. To accomplish this, two position (on/off) controllers (pneumatic or electronic) are used. Proper selection of set points for such controllers is important so that meter runs will not be *cycling* (i.e., switch back and forth). A method for determining set points for two-position controllers is given in this chapter.

Determining set point for two-point differential gap controllers

Many times automatic meter tubes are used in order to have greater rangeability. Set points on controllers will vary depending on the orifice plate sizes, and should be determined very accurately to avoid continuous cycling or too high or too low differential readings. This can be done in much the same way that we determine what our new differential would be after an orifice plate change, by the equation

$$h_2 = h_1 \left(\frac{C_1}{C_2}\right)^2 \tag{2.1}$$

In determining controller set points, however, the total of coefficients must be used. We can then say that

$$h_2 = h_1 \left(\frac{\text{total of coefficients before tube switch}}{\text{total of coefficients after tube switch}}\right)^2 \tag{2.2}$$

For example, let's assume that we have four 12-in meter tubes in parallel and tubes 2, 3, and 4 have automatic valves and two position controllers. If we have a 12 × 6-in orifice in the no. 1 tube, a 12 × 6.250-in orifice in the no. 2 tube, a 12 × 6.500-in orifice in the no. 3 tube, and a 12 × 6.750-in plate in the fourth tube, and we select a high trip point of 90 in on no. 1 tube, what is the maximum value for

TABLE 2.1   Orifice Meter Tube Capacity Chart

Tube size, in	Actual inside diameter, in	Max. orifice bore, in	Hourly basic orifice factor $F_b - G$ = 1.0	Measuring pressure, lb/in² gauge				
				100	200	300	400	500
2	2.067	1.250	345.1	0.818	1.126	1.373	1.588	1.783
	1.939	1.125	276.2	0.655	0.901	1.099	1.271	1.427
3	3.068	1.750	663.4	1.572	2.164	2.639	3.053	3.427
	2.900	1.750	663.4	1.598	2.200	2.683	3.103	3.484
4	4.026	2.375	1,231.7	2.918	4.018	4.900	5.667	6.364
	3.826	2.250	1,104.7	2.617	3.604	4.394	5.083	5.706
6	6.065	3.625	2,876.0	6.813	9.381	11.441	13.232	14.855
	5.761	3.500	2,695.1	6.385	8.791	10.721	12.400	13.920
8	7.981	4.750	4,928.1	11.674	16.074	19.603	22.673	25.453
	7.625	4.500	4,412.8	10.454	14.394	17.553	20.302	22.792
10	10.020	6.000	7,872.9	18.650	25.679	31.316	36.220	40.662
	9.562	5.750	7,240.4	17.151	23.616	28.800	33.310	37.395
12	11.938	7.000	10,649.0	25.225	34.732	42.356	48.989	54.997
	11.374	6.750	9,935.7	23.536	32.407	39.521	45.710	51.316
16	15.000	9.000	17,711.0	41.952	57.766	70.466	81.479	91.471
	14.312	8.500	15,720.0	37.313	51.378	62.656	72.469	81.356

Notes: Capacity in MMscf/day.
Pipe tap capacity = tabulated volume × 1.26. Flow conditions: flange taps, mercury meter, downstream, static, ΔP = 50 in, G = 0.600, T = 60°F, CO_2 = 0.5%, N_2 = 0.5%, P_b = 14.73 lb/in² absolute.

the low set point on no. 1 controller? The orifice coefficients are 8859.8, 9817.2, 10,860.0, and 11,998.0, respectively. We can calculate this as follows:

$$h_2 = 90\left(\frac{8859.8}{8859.8 + 9817.2}\right)^2 = 20.3 \text{ in}$$

Our *maximum* low set point would be 20.3 in minus a safety margin of about 4 in, or 16 in. The safety margin would vary, depending on the type of load we were dealing with and the amount of swing caused by the tube switch itself. Now we determine the low set point for the second controller. (It controls the third tube.) We have:

$$h_2 = 90\left(\frac{8859.8 + 9817.2}{8859.8 + 9817.2 + 10,860.0}\right)^2 = 35.9 \text{ in} \qquad (2.3)$$

TABLE 2.1   Orifice Meter Tube Capacity Chart (*Continued*)

Measuring pressure, lb/in² gauge

600	700	800	900	1000	1100	1200	1300	1400
1.964	2.134	2.296	2.450	2.601	2.745	2.886	3.022	3.154
1.571	1.707	1.837	1.961	2.081	2.197	2.309	2.418	2.524
3.774	4.101	4.412	4.711	4.998	5.277	5.546	5.808	6.062
2.837	4.169	4.485	4.789	5.081	5.364	5.638	5.905	6.163
7.006	7.613	8.191	8.745	9.279	9.796	10.297	10.783	11.255
6.284	6.828	7.346	7.843	8.322	8.785	9.235	9.671	10.094
16.360	17.777	19.126	20.420	21.667	22.874	24.043	25.178	26.280
15.330	16.658	17.923	19.135	20.303	21.435	22.531	23.594	24.627
28.031	30.460	32.771	34.988	37.126	39.193	41.397	43.142	45.030
25.101	27.257	29.345	31.330	33.244	35.095	36.890	38.631	40.322
44.781	48.661	52.354	55.895	59.310	62.613	65.814	68.921	71.938
41.183	44.751	48.148	51.405	54.545	57.582	60.527	63.384	66.159
60.568	65.816	70.811	75.601	80.219	84.687	89.017	93.219	97.299
56.514	61.410	66.071	70.540	74.849	79.018	83.058	86.979	90.786
100.737	109.464	117.772	125.740	133.421	140.851	148.053	155.042	161.829
89.597	97.359	104.749	111.835	118.667	125.275	131.681	137.897	143.934

To change to other flow conditions:

$$Q = \text{(tabulated vol.)} \sqrt{\frac{0.6}{G_{\text{flow}}}} \sqrt{\frac{520}{T_{\text{flow}}}} \left[\frac{14.73}{P_b(\text{new})}\right] \sqrt{\frac{\Delta P_{\text{flow}}}{50}} \quad \text{(use only factor} \atop \text{where change occurs)}$$

Volume at 90 in = tabulated volume × 1.342.
Orifice sizes selected to make β = 0.600 as nearly as possible.
Use 80% of chart volumes for load factor.

If we allow a safety margin of 4 in, our set point would be 31.9 in. The low set point for the third controller (fourth tube) would be

$$h_2 = 90 \left(\frac{8859.8 + 9817.2 + 10{,}860.0}{8859.8 + 9817.2 + 10{,}860.0 + 11{,}998.0}\right)^2 = 45.5 \text{ in} \qquad (2.4)$$

minus a safety margin.

If it is desirable to select a low set point and calculate the high, this is done in exactly the same manner, but the safety margin must be added rather than subtracted.

It must also be remembered that these calculations will hold true only if the meter tubes are well-balanced.

TABLE 2.2 Multipliers for Changing Orifices

Average differential reading	Desired differential reading									
	0.2	0.5	1.0	2.0	5.0	10	15	20	40	60
0.1	0.85	0.68	0.57	0.50	0.46	0.40				
0.2	1.00	0.80	0.68	0.57	0.57	0.50		0.41		
0.5	1.25	1.00	0.85	0.72	0.58	0.57	0.45	0.50	0.41	
1.0	1.47	1.18	1.00	0.85	0.80	0.68	0.53	0.57	0.50	0.45
2.0	1.75	1.40	1.18	1.00	1.00	0.85	0.62	0.72	0.62	0.55
5.0		1.75	1.47	1.25	1.18	1.00	0.78	0.85	0.72	0.65
10			1.75	1.47	1.40	1.18	0.95	1.00	0.85	0.68
20				1.75	1.65	1.40	1.10	1.18	1.00	0.92
40					1.95	1.65	1.27	1.40	1.18	1.08
80							1.50			

Example: The average differential on a 50-in orifice meter for a 3-in orifice is 2 in. It is desired to obtain an average reading of 20 in. The multiplier is 0.57. (Old plate size × multiplier = 3 × 0.57 = 1.71.) Use 1¾-in orifice.

2.3   Operation and Maintenance of
Turbine Meters

General

The operation of a turbine meter is such that any change in accuracy will be detected by visually inspecting the internal mechanism, by spin testing, or by checking its accuracy near the minimum recommended flow rates. Obviously, if the rotor has been damaged by foreign substances or objects going through the meter and breaking the rotor blades, then the meter will be inaccurate at all flow rates. Similarly, if some object blocked off the passages so that the gas could flow only through a small section of the meter, the inaccuracy would occur at both high and low rates of flow. Normally, meter accuracy changes are due to frictional drag starting at the low capacity end of the meter. It is, therefore, important to have very little frictional drag in the meter and also to keep the energy consumed by the index and/or recording instrument to a minimum.

Spin test

Since friction is the only nonvisual factor which will cause a change in calibration, a test which will accurately indicate any change in friction is a test for meter accuracy. The spin time test determines the relative level of mechanical friction in the meter and is one of the most commonly applied field checks. If the mechanical friction has not significantly changed, the meter area is clean, and the internal portions of the meter show no damage, the meter should display no change in accuracy. If the mechanical friction has increased significantly, this indicates that the accuracy characteristics of the meter at low base flow rates has been degraded. Figure 2.1 demonstrates the typical affect of mechanical friction on the accuracy of a turbine meter. The normal procedure for making a spin test is to remove the module from the meter.

The test must be made in a draft-free environment and the mechanism must be held in its normal operating position. The rotor is set into rotation by a quick twist of the finger and the rotor is timed from the initial rotation until the rotor stops. Spin times should be repeated at least three times and the average time determined and recorded in order to compare it with the manufacturer's recommended spin time. Spinning the rotor by hand will result in spin times which are repeatable within 2 to 3 seconds.

It is recommended that the first spin test be conducted with the internal mechanism completely assembled except for index or recording gauges. This first average spin time should be noted. Not only will

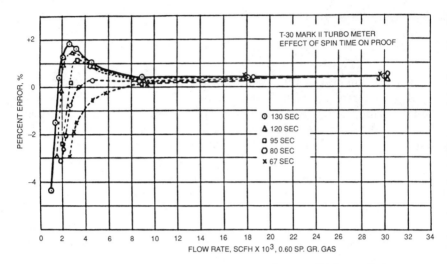

Figure 2.1   Low-flow accuracy at various spin times. (*Courtesy Equimeter Co.*)

this check the main rotor bearings but it will also check the other bearings and gearing for friction. If the spin times are appreciably (25 to 30 percent) less than those recommended by the manufacturer, the tests should be repeated at various levels of disassembly and noted until the source of the abnormal friction is located. The usual cause of a change in spin time is increased rotor shaft bearing friction. Cleaning and oiling the bearings usually brings the spin time back to normal; however, it should be noted that there are other points in the mechanism where mechanical friction can exist and affect spin time. The spin times for various stages of disassembly will usually reveal the problem area. Then the module, index, and gauge can be reassembled. Spin tests should then be made and times noted.

When future spin tests are made, the "as-found" spin times should be compared to the initial tests. If later "as-found" spin time tests with module, index, and gauge complete are not appreciably less than initial spin time tests, no further module teardown is needed. Then all that is required is to clean and oil the module, complete the "as-left" spin test, and reinstall the module. It is not a cause for alarm if sometimes the as-left spin test measurement is somewhat lower, since overoiling will temporarily reduce spin time.

One should bear in mind that spin tests are also preventive maintenance tests. As a rule, progressive deterioration of spin time is a good indication of future mechanical problems. If cleaning and flushing of bearings do not improve spin times and smoothness of rotation, then bearings should be replaced.

Visual inspection

Routine field inspection of turbine meters is necessary in order to assure accurate measurement. Because of the straight-through flow path, the small number of moving parts, and the ready accessibility to meter internals, maintenance is quite simple.

Meters which are operating can often yield information by the noise they generate or by vibrations felt through the body. If the meter has severe vibration it usually indicates damage which has unbalanced the rotor, and this condition will lead to complete rotor failure. Rotor rubbing and poor bearings can often be heard and/or felt at relatively low flow rates, where such noises are not masked by normal flow noise.

An internal inspection can be accomplished by removing the module from the meter body. If the module is not removable, the entire meter may be removed from the run, or closures on the run may be removed for internal inspection.

During visual inspection, the rotor should be examined for missing blades and erosion or other damage that would affect rotor balance and blade configuration. Dirt or foreign material must not be permitted to accumulate on or around the rotor blades or on the module and nose cone. Rotor clearances should be checked to ensure equal spacing on both sides of the rotor where it protrudes into the housing.

If the rotor is damaged in any way, the mechanism should be replaced with a precalibrated mechanism, or the entire meter should be changed.

Cleaning and lubrication

It is recommended that, when a meter is disassembled for any purpose, the mechanism be thoroughly cleaned to remove dirt or foreign material. Bearings should be oiled in accordance with the manufacturer's recommendation. Careful attention should be given to areas where excessive lubricant would cause a liquid drag.

The two main bearings operate at the highest rotational speed, and, because of this, many turbine meters have provisions for externally oiling the rotor shaft bearings. When the meter is disassembled, oil should be added through the outside lubricator and a visual check made to be certain that lubricant is flowing freely to both main bearings. While the module is out of the meter, it is easy to check the exposed screws and nuts to be certain they are still tight.

How frequently a meter should be lubricated while it is in service depends on operating conditions and the condition of the gas. Meters operating at high flow rates and with dirty gas may require weekly lubrication. Meters operating at average flow rates and on relatively clean gas may require lubrication only on a monthly basis. It is a good procedure to require that chart changers inject several drops of turbine lubricant in the external oiling mechanism at each chart changing interval (usually weekly). With frequent oiling, the bearings are periodically flushed of grit and dirt, and thus bearing life and meter accuracy are preserved. An overoiled meter presents only a temporary slowdown of the rotor, whereas dry bearings are conducive to rotor bearing deterioration and meter stoppage.

New installation

A turbine module should never be left in the body during any hydro or gas pressure test. Prior to such tests, the module should be removed from meter body and top plate. Then the top plate can be replaced for the tests. The entire meter piping should be purged, meter body cleaned, and module replaced.

Reasonability tests

To detect any installation errors, a *reasonability test* should be conducted. The test should be repeated every 6 months or after any change of module, index, change gears, recording instrument, or integrating gauge. The reasonability test should consist of one low flow rate test with a critical flow prover or any other series testing device. If the test shows an error in excess of 2 percent, the module should be removed as soon as possible and returned to a test site or manufacturer for retest.

Filters and strainers

One of the most significant and frequent causes of bearing wear and module failure is from trash or foreign material in the pipeline flow stream if the meter is not properly protected. There are times when, because of peak load demands or plant problems, the normal pipeline gas quality standards are sacrificed for one reason or another. This pipeline quality upset is caused by new tie-ins; hot taps; normal internal corrosion in the form of dust, dirt, and scale; welding beads; and welding rods. For this reason, it is important that filters or strainers be installed upstream of the meter installation. In some stations, strainers with wire mesh are installed to catch most of this trash. In some instances, it may be preferable to install filters that remove fine dust to increase bearing life.

Normally some type of differential pressure gauge is installed so that the pressure drop across the filter or strainer can indicate the excessive pressure drop from a buildup of foreign matter in the filter or strainer. Normal pressure drop should be observed initially and at various flow rates when the strainer or filter is clean. The devices should be inspected whenever excessive pressure drops are indicated on the differential pressure device.

All foreign material inside the meter approach piping should be carefully removed before the meter module is installed.

Sealed bearings

Sometimes, because of high volume and pressure conditions and the bad quality of the gas, sealed bearings should be used. Sealed bearings look somewhat like shielded bearings except that both faces are sealed to the inner and outer races and a permanent lubricant is packed between the bearing faces. The reason for sealing the faces is to prevent solids as well as liquids from entering the bearings. Sometimes sealed bearings are made larger than semishielded or shielded bearings. This increase in size is required for longer life with high load conditions. Because of the sealing of bearing faces and the packing of the bearings with lubricant, initially the spin times will usually be low in sealed bearings. But after the sealed bearings have been in use for a few weeks, the spin times should increase to at least the manufacturer's published minimum time for a sealed bearing. The increase of spin time is a result of the "wearing-in" of the bearings, races, and seals, and the driving out of excess packed lubricant by centrifugal force.

At the initial spin test with any new sealed bearings, the time should be noted and referred to on any subsequent tests. These spin times should increase dramatically after the first few months and then level off. When subsequent spin tests are conducted, careful note should be made as to whether the bearings are noisy, or quiet and smooth. Also, attention should be paid to radial (side to side) or longitudinal (front and back) slack of bearings on the rotor shaft. If any of the above—noise or radial or longitudinal slack—is observed, old bearings should be replaced as soon as possible, regardless of spin times. If bearings show slow times after their wear-in period, they should be replaced as soon as possible. New bearings should be pur-' chased according to the manufacturer's recommendation.

Sometimes, after sealed bearing modules have been installed and placed in service, the pressure, gas quality, or volume conditions no longer require sealed bearings. Then the sealed bearing module should be replaced with standard bearings, or sealed bearings should be replaced with standard bearings in the existing module. This replacement is especially required when the metered volume declines to a very low value (below 10 percent of full flow). The rangeability of a turbine meter with large, heavy-duty sealed bearings is about one-half the rangeability of a standard bearing module. Hence, continued low-volume service should be avoided with sealed bearings.

2.4 Prevention of Gas Hydrate Formation

Gas hydrate is a snowy or icelike combination of solid and water. Continued cooling of natural gas with entrained moisture can result in gas hydrate, causing serious damage to equipment and stopping gas flow. Gas hydrate, as usually obtained from a pipeline, looks like packed snow. For hydrate formation, water must be present and the gas stream must be at or near the hydrate formation temperature at system pressure.

The most effective way to prevent hydrate formation is to prevent formation of conditions causing hydrates. One major factor is water. Dehydrating the gas is the best solution. Another approach is heating the gas above hydrate temperature. This is an expensive method, although line heaters are used in many places. A third solution is to lower system pressure. This is not always acceptable, since lowering pressure also means lowering system capacity. A fourth method is inhibitor injection. Injecting methanol is thought to be one of the best methods for preventing gas hydrate formation. Methanol will work to decompose blockages and to prevent their formation. Table 2.3 shows the rate at which methanol should be injected to prevent hydrates. Table 2.4 shows moisture content in air at various dew points.

2.5 Errors in Orifice Measurement

Any measurement is associated with inaccuracy or uncertainty. Measurement of gaseous fluid is not an exact science. With inferential-type measurements such as orifice metering, errors can be introduced by various contributing factors, some of which are difficult to control. Installation condition, operating practices, and the quality of gas being measured can add to the inherent uncertainty of the device. Tables 2.5 to 2.7 were generated during studies and tests conducted at facilities of Florida Gas Transmission Co. and Conoco. Table 2.8 shows percent error in orifice metering produced by static or differential error that can be caused by equipment being out of calibration. The United Gas Pipeline Co. also conducted some tests with the orifice plate inserted backward, i.e., with the bevel facing upstream. The amount of correction to the measured volume (the adjustment) is then plotted against beta ratio (β)(Fig. 2.2). In all of these instances, measured volumes appear to be lower than the actual volumes.

With no formula or analytical methods, such test results may be

TABLE 2.3   Rates to Feed Alcohol or Methanol to Prevent Hydrates

Gas volume per day, Mscf	Gallons per day	Drops per day	Drops per minute
25	0.025	2,555	2
50	0.050	5,110	4
100	0.100	10,220	7
125	0.125	12,774	9
150	0.150	15,329	11
200	0.200	20,439	14
225	0.225	22,994	16
250	0.250	25,549	18
275	0.275	28,104	20
300	0.300	30,659	21
325	0.325	33,213	23
350	0.350	35,768	25
375	0.375	38,323	27
400	0.400	40,878	28
425	0.425	43,433	30
450	0.450	45,988	32
475	0.475	48,543	34
500	0.500	51,097	35
550	0.550	56,207	39
600	0.600	61,317	43
650	0.650	66,427	46
700	0.700	71,537	50
750	0.750	76,646	53
800	0.800	81,756	57
850	0.850	86,866	60
900	0.900	91,976	64
1000	1.000	102,195	71
1125	1.125	114,969	79
1250	1.250	127,743	89
1375	1.375	140,518	98
1500	1.500	153,293	106
1625	1.625	166,067	115
1750	1.750	178,841	124
1875	1.875	191,116	133
2000	2.000	204,390	142

102,195 drops = 1 gallon
6,387 drops = ½ pint
12,774 drops = 1 pint
25,549 drops = 1 quart

used for making volumetric (accounting) adjustment. These results do not cover all different sizes or every possible error; however, they do provide an idea of the magnitude of a problem. Measurement error due to bending (buckling) of the orifice plate was studied by D. Mason, M. P. Wilson, Jr., and W. G. Birkhead, who found error to be as great as 0.62 percent for a 6 × 1.8-in orifice. The graphs and methods to calculate error due to bending are those suggested by Mason, Wilson, and

TABLE 2.4 Moisture Content of Air

Dew point, °F	Dew point, °C	ppm† by volume	ppm† by weight	Pounds water per MMscf dry air
-110	-78.9	0.63	0.39	0.032
-109	-78.3	0.69	0.43	0.034
-108	-77.8	0.75	0.47	0.038
-107	-77.2	0.82	0.51	0.041
-106	-76.7	0.88	0.55	0.044
-105	-76.1	1.00	0.62	0.050
-104	-75.5	1.08	0.67	0.054
-103	-75.0	1.18	0.73	0.059
-102	-74.4	1.29	0.80	0.065
-101	-73.9	1.40	0.87	0.070
-100	-73.3	1.53	0.95	0.077
-99	-72.8	1.66	1.03	0.083
-98	-72.2	1.81	1.12	0.091
-97	-71.7	1.96	1.22	0.098
-96	-71.1	2.15	1.34	0.108
-95	-70.6	2.35	1.46	0.118
-94	-70	2.54	1.58	0.127
-93	-69.4	2.76	1.72	0.138
-92	-68.9	3.00	1.86	0.150
-91	-68.3	3.28	2.04	0.164
-90	-67.8	3.53	2.19	0.177
-89	-67.2	3.84	2.39	0.192
-88	-66.7	4.15	2.58	0.221
-87	-66.1	4.50	2.80	0.225
-86	-65.6	4.78	2.97	0.239
-85	-65.0	5.30	3.30	0.265
-84	-64.4	5.70	3.50	0.285
-83	-63.9	6.2	3.9	0.31
-82	-63.3	6.6	4.1	0.33
-81	-62.8	7.2	4.5	0.36
-80	-62.2	7.8	4.8	0.39
-79	-61.7	8.4	5.2	0.42
-78	-61.1	9.1	5.7	0.46
-77	-60.6	9.8	6.1	0.49
-76	-60.0	10.5	6.5	0.53
-75	-59.4	11.4	7.1	0.57
-74	-58.9	12.3	7.6	0.62
-73	-58.3	13.3	8.3	0.67
-72	-57.9	14.3	8.9	0.72
-71	-57.2	15.4	9.6	0.77
-70	-56.7	16.6	10.3	0.83
-69	-56.1	17.9	11.1	0.96
-68	-55.6	19.2	11.9	1.0
-67	-55.0	20.6	12.8	1.03
-66	-54.4	22.1	13.7	1.1
-65	-53.9	23.6	14.7	1.2
-64	-53.3	25.6	15.9	1.3
-63	-52.8	27.5	17.1	1.4
-62	-52.2	29.4	18.3	1.5
-61	-51.7	31.7	19.7	1.6
-60	-51.1	34.0	21.1	1.7
-59	-50.6	36.5	22.7	1.8
-58	-50.0	39.0	24.2	2.0
-57	-49.4	41.8	26.0	2.1
-56	-48.9	44.6	27.7	2.2
-55	-48.3	48.0	30	2.4
-54	-47.8	51	32	2.6
-53	-47.2	55	34	2.8
-52	-46.7	59	37	2.9
-51	-46.1	62	39	3.1
-50	-45.6	67	42	3.4
-49	-45.0	72	45	3.6
-48	-44.4	76	47	3.8
-47	-43.9	82	51	4.1
-46	-43.3	87	54	4.4
-45	-42.8	92	57	4.6
-44	-42.2	98	61	4.9
-43	-41.7	105	65	5.3
-42	-41.1	113	70	5.7
-41	-40.6	119	74	6.0
-40	-40.0	125	80	6.4
-39	-39.4	136	85	6.8
-38	-38.9	144	89	7.2
-37	-38.3	153	95	7.7
-36	-37.9	164	102	8.2

-35	-37.2	174	108	8.7
-34	-36.7	185	115	9.3
-33	-36.1	196	122	9.8
-32	-35.6	210	131	10.5
-31	-35.0	222	140	11.1
-30	-34.4	235	146	11.8
-29	-33.9	250	155	12.5
-28	-33.3	265	165	13.3
-27	-32.8	283	176	14.2
-26	-32.2	300	186	15.0
-25	-31.7	317	197	15.9
-24	-31.1	338	210	16.9
-23	-30.6	358	223	17.9
-22	-30.0	378	235	18.9
-21	-28.4	400	249	20.0
-20	-28.9	422	212	21.1
-19	-28.3	448	278	22.4
-18	-27.8	475	295	23.8
-17	-27.2	500	311	25.0
-16	-26.7	530	329	26.5
-15	-26.1	560	348	28.0
-14	-25.6	590	367	29.5
-13	-25.0	630	391	31.5
-12	-24.4	660	410	33.0
-11	-23.9	700	435	35.0

-10	-22.3	740	460	37.0
-9	-22.8	780	485	39.0
-8	-22.2	820	509	41.0
-7	-21.7	870	541	43.5
-6	-21.1	920	572	46.0
-5	-20.6	970	603	48.5
-4	-20.0	1020	634	51.0
-3	-19.4	1087	675	54.4
-2	-18.9	1153	716	57.7
-1	-18.3	1220	758	61.0
0	-17.8	1299	807	37.0
1	-17.2	1378	856	39.0
2	-16.7	1458	906	41.0
3	-16.1	1537	955	43.5
4	-15.6	1616	1004	46.0
5	-15.0	1695	1053	37.0
6	-14.4	1775	1103	39.0
7	-13.9	1854	1152	41.0
8	-13.3	1933	1201	43.5
9	-12.8	2053	1276	46.0
10	-12.2	2174	1351	109
11	-11.7	2294	1425	115
12	-11.1	2415	1501	121
13	-10.6	2535	1575	127
14	-10.0	1656	1650	133

15	-9.4	2776	1725	139
16	-8.9	2897	1800	145
17	-8.3	3017	1875	151
18	-7.8	3197	1986	160
19	-7.2	3376	2098	169
20	-6.7	3556	2209	178
21	-6.1	3735	2321	187
22	-5.6	3915	2432	196
23	-5.0	4094	2544	205
24	-4.4	4274	2656	214
25	-3.9	4453	2767	223
26	-3.3	4633	2879	232
27	-2.8	4883	3034	244
28	-2.2	5133	3189	357
29	-1.7	5383	3345	269
30	-1.1	5633	3400	282
31	-0.56	5883	3655	294
32	0	6133	3811	307
33	0.56	6383	3966	319
34	1.1	6633	4121	332
35	1.7	6883	4277	344
36	2.2	7281	4524	364
37	2.8	7679	4771	384
38	3.3	8077	5019	404
39	3.9	8475	5266	424

TABLE 2.4 Moisture Content of Air (Continued)

Dew point, °F	Dew point, °C	ppm† by volume	ppm† by weight	Pounds water per MMscf dry air
40	4.4	8,874	5,514	444
41	5.0	9,272	5,761	464
42	5.6	9,670	6,008	484
43	6.1	10,068	6,256	503
44	6.7	10,466	6,503	522
45	7.2	10,824	6,725	541
46	7.8	11,182	6,948	559
47	8.3	11,540	7,170	577
48	8.9	11,898	7,393	595
49	9.4	12,255	7,614	613
50	10.0	12,613	7,837	631
51	10.6	12,971	8,059	649
52	11.1	13,329	8,282	666
53	11.7	13,687	8,504	684
54	12.2	14,268	8,865	713
55	12.8	14,849	9,226	742
56	13.3	15,430	9,587	772
57	13.9	16,011	9,948	801
58	14.4	16,593	10,310	830
59	15.0	17,174	10,671	859
60	15.6	17,754	11,031	889
61	16.1	18,336	11,393	917
62	16.7	18,917	11,754	946
63	17.2	19,685	12,231	984
64	17.8	20,452	12,707	1023
65	18.3	21,220	13,185	1061
66	18.9	21,987	13,661	1099
67	19.4	22,755	14,138	1138
68	20.0	23,522	14,615	1176
69	20.6	24,290	15,092	1215
70	21.1	25,057	15,569	1253
71	21.7	25,825	16,046	1291
72	22.2	26,827	16,668	1341
73	22.8	27,829	17,291	1391
74	23.3	28,831	17,914	1442
75	23.9	29,883	18,536	1492
76	24.4	30,836	19,159	1542
77	25.0	31,838	19,782	1592
78	25.6	32,840	20,405	1642
79	26.1	33,842	21,027	1692
80	26.7	34,844	21,650	1742
81	27.2	36,138	22,454	1807
82	27.8	37,432	23,258	1872
83	28.3	38,727	24,062	1936
84	28.9	40,021	24,866	2001
85	29.4	41,315	25,670	2066
86	30.0	42,609	26,474	2130
87	30.6	43,903	27,278	2195
88	31.1	45,198	28,083	2260
89	31.7	46,492	28,887	2325
90	32.2	48,146	29,914	2407
91	32.8	49,801	30,943	2490
92	33.3	51,455	31,971	2573
93	33.9	53,109	32,998	2655
94	34.4	54,764	34,027	2738
95	35.0	56,419	35,055	2821
96	35.6	58,023	36,083	2904
97	36.1	59,728	37,111	2986
98	36.7	61,382	38,139	3069
99	37.2	63,477	39,440	3174
100	37.9	65,571	40,741	3279
101	38.3	67,666	42,043	3383
102	38.9	69,760	43,344	3488
103	39.4	71,855	44,646	3593
104	40.0	73,950	45,948	3698
105	40.6	76,045	47,249	3802
106	41.1	78,139	48,550	3907
107	41.7	80,234	49,852	4012
108	42.2	83,861	51,484	4193
109	42.8	85,489	53,117	4274
110	43.3	88,116	54,749	4406

*At 14.7 lb/in^2 absolute
†Parts per million

TABLE 2.5   Errors in Measurement

Condition	% difference
Orifice edge beveled 45° full circumference (machined):	
0.010-in bevel width	– 2.2
0.020-in bevel width	– 4.5
0.050-in bevel width	– 13.1
Turbulent gas stream:	
Upstream valve partially closed—straightening vanes in	– 0.7
Upstream valve partially closed—straightening vanes out	– 6.7
Liquid in bottom of meter tube, 1-in deep	– 11.3
Grease and dirt deposits in meter tube	– 11.1
Leaks around orifice plate:	
One clean cut through plate sealing unit	
a. Cut on top side of plate	– 3.3
b. Cut next to tap holes	– 6.1
Orifice plate carrier raised approximately ⅜ in from bottom	– 8.2
(plate not centered)	
Valve lubricant on upstream side of plate:	
Bottom half of plate coated ¹⁄₁₆ in thick	– 9.7
Three gob-type random deposits	0
Nine gob-type random deposits	– 0.6
Orifice plate uniformly coated ¹⁄₁₆ in over full face	– 15.8
Valve lubricant on both sides of plate:	
Plate coated ⅛ in both sides full face	– 17.9
Plate coated ¼ in both sides full face	– 24.4
Plate coated ⅛ in bottom half both sides	– 10.1
Plate warp tests:	
Plate warped toward gas flow ⅛ in from flat	– 2.8
Plate warped away from gas flow ⅛ in from flat	– 0.6
Plate warped toward gas flow ¼ in from flat	– 9.1
Plate warped away from gas flow ¼ in from flat	– 6.1

Birkhead (Figs. 2.3 to 2.6). These results are furnished for information only. This author or the parties involved in these studies and publication thereof cannot be held responsible for the accuracy and the use of this information.

Square root error in chart integration process

One of the common practices in integrating a bad chart is to draw a line through the paint stripe which is then used for the actual integration process. If this line is drawn through the middle of a pulsating differential (i.e., if the average differential is accepted for integration), it will invariably result in a higher flow rate than actual. The percent error will depend on the width of the paint stripe or amplitude of pul-

TABLE 2.6   Common Errors in Gas Measurement Caused by Between-the-Taps Mechanical Defects (Gremlins) in 2- through 6-in Orifice Meter Runs

Gremlin	Cause	Effect on measurement accuracy
6-in plate bowed downstream ¼ in	Upstream freeze-up plate bowed by gas pressure or ice slug	Up to 6%
Upstream solids or liquids resting against plate (same condition with wrong-schedule donut seal in orifice fitting)	Accumulation of mud, water hydrates, pipe junk, welding gloves	Up to 16%
Beveled-edge plate installed with bevel facing upstream	Carelessness	Up to 15%, depending on orifice size
Badly worn plate with upstream edge of orifice bore slightly rounded	Sand, solids, time	Up to 4.5%
Dull plate—looks sharp but won't pare fingernail	Wear	Up to 2%
Nicks	Dropped, mechanical damage	Hard to classify, but 1 to 2% is common
Leaky seal ring around plate, but ring continuous	Deteriorated rubber, O ring, or cut edges	Up to 2%
Discontinuous or cut out seal ring	Damaged on installation or removal	3% +
An orifice plate stamped 4 × 1.000 in but actually bored ¹⁄₃₂ in oversize	Accidentally (or intentionally) misbored	6.7%

sation. Such an error is called a *square root error* (Fig. 2.7). Assuming there is no other error, this square root error is simply a data processing error which can be avoided or minimized by adopting a differential pressure line for integration that will represent a more accurate flow rate.

This can be achieved by using the following relation to arrive at a differential pressure for integration purpose:

$$\Delta P_o \text{ (for integration)} = \left(\frac{\sqrt{\Delta p_{\max}} + \sqrt{\Delta p_{\min}}}{2} \right)^2 \tag{2.5}$$

Example

$$\Delta p_{\text{avg}} = 50\% \text{ on chart}$$

Pulsation (or paint width) = 40% (i.e., ±20% from mean)

TABLE 2.7   Effect of Orifice Plate Fouling on Measurement Accuracy ($\beta = 0.2$)

Type of fouling	Flow measurement error, %
Sand	
1 sand quadrant	−0.97
2 sand quadrants	−2.79
3 sand quadrants	−3.91
4 sand quadrants	−6.22
4 sand quadrants with 6-mm ring removed from around orifice bore	−0.31
Grease	
4 grease deposits	−1.02
8 grease deposits	−2.75
16 grease deposits	−2.14
32 grease deposits	−2.57

TABLE 2.8   Percent Error in Orifice Metering Produced by Static or Differential Error

Error, lb/in² high	Gauge pressure, lb/in²								
	25	50	100	150	200	250	300	350	400
1	1.25	0.75	0.45	0.30	0.25	0.20	0.16	0.14	0.12
2	2.50	1.55	0.85	0.60	0.45	0.40	0.32	0.28	0.24
3	3.70	2.30	1.30	0.90	0.70	0.55	0.48	0.42	0.36
4	4.90	3.05	1.75	1.20	0.95	0.75	0.64	0.56	0.48
5	6.10	3.80	2.15	1.50	1.15	0.95	0.80	0.70	0.60

Error, inH₂O high	Differential pressure, inH₂O								
	10	20	30	40	50	60	70	80	90
0.1	0.50	0.25	0.17	0.12	0.10	0.08	0.07	0.06	0.06
0.2	1.00	0.50	0.33	0.25	0.20	0.17	0.14	0.13	0.11
0.3	1.50	0.75	0.50	0.37	0.30	0.25	0.21	0.19	0.17
0.4	2.00	1.00	0.66	0.50	0.40	0.33	0.28	0.25	0.22
0.5	2.45	1.25	0.83	0.62	0.50	0.42	0.36	0.31	0.28
0.6	2.95	1.50	1.00	0.75	0.60	0.50	0.43	0.38	0.33
0.7	3.45	1.75	1.16	0.87	0.70	0.58	0.50	0.44	0.39
0.8	3.90	2.00	1.32	1.00	0.80	0.66	0.57	0.50	0.44
0.9	4.40	2.25	1.49	1.12	0.90	0.75	0.64	0.56	0.50
1.0	4.90	2.50	1.65	1.24	1.00	0.82	0.71	0.62	0.55

Note: Percent error produced by low static or differential pressure is practically the same as that shown for high static or differential.

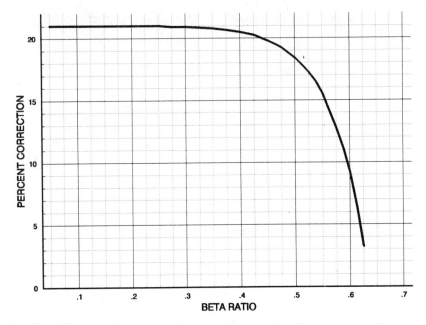

Figure 2.2   Correction required for orifice plate with 45° bevel on upstream side.

$$\Delta p_{\max} = 70\%$$

$$\Delta p_{\min} = 30\%$$

$$\Delta p_o = 47.91\% \text{ (approx. 48\%)}$$

$$\% \text{ error in flow rate} = \frac{\sqrt{\Delta p_{\text{avg}}} - \sqrt{\Delta P_o}}{\sqrt{\Delta P_o}} \times 100 = 2\%$$

Note that this calculation needs to be done only once as long as the paint width or pulse amplitude remains the same. However, for each significant change in pulse amplitude or paint width from the average, a new calculation should be done and a new line drawn for the respective period. These lines can then be joined to provide a continuous line for integration purposes.

2.6   Meter Accuracy and Correction

The percentage of error denotes the variation from the actual value determined by using an accepted standard. The percentage of correction, on the other hand, reflects the amount of adjustment to be made on the erroneous reading by the meter in service. In mathematical form,

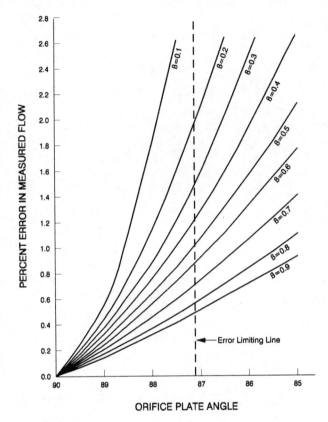

Figure 2.3 Measurement error vs. orifice plate angle. (*Courtesy W. G. Birkhead*)

$$\% \text{ error} = \frac{(\text{meter reading} - \text{prover reading})}{\text{prover reading}} \times 100 \qquad (2.6)$$

$$\% \text{ correction} = \frac{(\text{meter reading} - \text{prover reading})}{\text{meter reading}} \times 100 \qquad (2.7)$$

A positive error or correction denotes that the meter is fast, i.e., the meter reading is higher than actual. Therefore, a positive correction should be subtracted from and a negative correction should be added to the meter reading to obtain the correct reading. Therefore,

$$\text{Adjustment} = - \text{meter reading} \times \% \text{ correction} \qquad (2.8)$$

If % correction is negative, the adjustment will be positive, and if % correction is positive, the adjustment will be negative.

If meter proving is done using the elapsed time during a test, and a

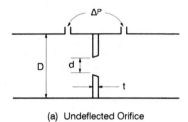

(a) Undeflected Orifice

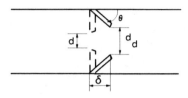

(b) Deformed Orifice

$$\theta = 90^\circ - \tan^{-1}\left[\frac{2\delta}{D\,(1-\beta)}\right]$$

Figure 2.4 Orifice plate deforma-
tion. (*Courtesy W. G. Birkhead*)

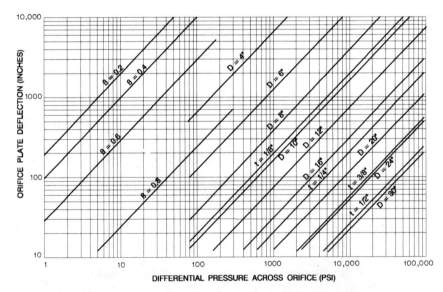

DIFFERENTIAL PRESSURE ACROSS ORIFICE (PSI)

Figure 2.5   Orifice plate deflection in flange fittings. (*Courtesy W. G. Birkhead*)

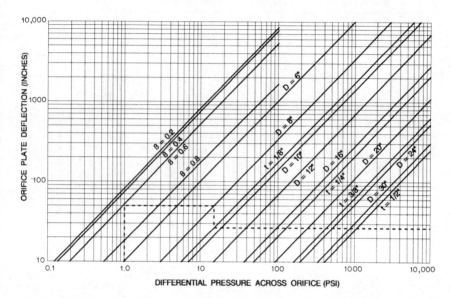

Figure 2.6   Orifice plate deflection in orifice fittings. (*Courtesy W. G. Birkhead*)

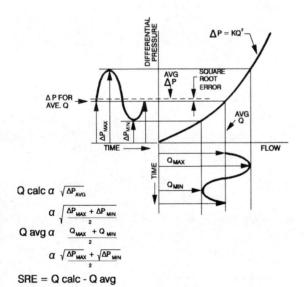

$$Q \text{ calc } \alpha \ \sqrt{\Delta P_{AVG}}$$

$$\alpha \ \sqrt{\frac{\Delta P_{MAX} + \Delta P_{MIN}}{2}}$$

$$Q \text{ avg } \alpha \ \frac{Q_{MAX} + Q_{MIN}}{2}$$

$$\alpha \ \frac{\sqrt{\Delta P_{MAX}} + \sqrt{\Delta P_{MIN}}}{2}$$

SRE = Q calc - Q avg

Figure 2.7   Illustration of square root error in orifice metering.

TABLE 2.9 Changing Percent Error to Percent Correction

% error in meter	% correction (or % error in gas bill)									
	0	0.1	0.2	0.3	0.4	0.5	0.6	0.7	0.8	0.9
	Fast meters									
0	0.0	0.1	0.2	0.3	0.4	0.5	0.6	0.7	0.8	0.9
1	1.0	1.1	1.2	1.3	1.4	1.5	1.6	1.7	1.8	1.9
2	2.0	2.1	2.2	2.2	2.4	2.4	2.5	2.6	2.7	2.8
3	2.9	3.0	3.1	3.2	3.3	3.4	3.5	3.6	3.7	3.8
4	3.8	3.9	4.0	4.1	4.2	4.3	4.4	4.5	4.6	4.7
5	4.8	4.9	5.0	5.0	5.1	5.2	5.3	5.4	5.5	5.6
6	5.7	5.8	5.8	5.9	6.0	6.1	6.2	6.3	6.4	6.5
7	6.6	6.6	6.7	6.8	6.9	7.0	7.1	7.2	7.2	7.3
8	7.4	7.5	7.6	7.7	7.8	7.8	7.9	8.0	8.1	8.2
9	8.3	8.4	8.4	8.5	8.6	8.7	8.8	8.8	8.9	9.0
10	9.1	9.2	9.3	9.3	9.4	9.5	9.6	9.7	9.8	9.8
11	9.9	10.0	10.1	10.2	10.2	10.3	10.4	10.5	10.6	10.6
12	10.7	10.8	10.9	11.0	11.0	11.1	11.2	11.3	11.4	11.4
13	11.5	11.6	11.7	11.7	11.8	11.9	12.0	12.0	12.1	12.2
14	12.3	12.4	12.4	12.5	12.6	12.7	12.7	12.8	12.9	13.0
15	13.0	13.1	13.2	13.3	13.4	13.4	13.5	13.6	13.6	13.7
16	13.8	13.9	14.0	14.0	14.1	14.2	14.2	14.3	14.4	14.5
17	14.5	14.6	14.7	14.7	14.8	14.9	15.0	15.0	15.1	15.2
18	15.3	15.3	15.4	15.5	15.6	15.6	15.7	15.8	15.8	15.9
19	16.0	16.0	16.1	16.2	16.2	16.3	16.4	16.5	16.5	16.6
20	16.7	16.7	16.8	16.9	17.0	17.0	17.1	17.2	17.2	17.3
21	17.4	17.4	17.5	17.6	17.6	17.7	17.8	17.8	17.9	18.0
22	18.0	18.1	18.2	18.2	18.3	18.4	18.4	18.5	18.6	18.6
23	18.7	18.8	18.8	18.9	19.0	19.0	19.1	19.2	19.2	19.3
24	19.4	19.4	19.5	19.6	19.6	19.7	19.8	19.8	19.9	19.9
25	20.0	20.1	20.1	20.2	20.3	20.3	20.4	20.4	20.5	20.6

standard time is given for the meter, the % error and % correction can be expressed in terms of time:

$$\% \text{ error} = \frac{\text{standard time} - \text{stopwatch time}}{\text{stopwatch time}} \times 100 \qquad (2.9)$$

$$\% \text{ correction} = \frac{\text{standard time} - \text{stopwatch time}}{\text{standard time}} \times 100 \qquad (2.10)$$

Table 2.9 provides relationships between % error and % correction for fast and slow meters.

Example

$$\text{Meter reading} = 1000 \text{ Mscf}$$

$$\text{Meter error} = 4.2\% \text{ fast}$$

From the table, meter correction required = 4%

$$\text{Therefore, adjustment} = -1000 \times 0.04 \text{ Mscf} = -40$$

TABLE 2.9   Changing Percent Error to Percent Correction (*Continued*)

% error in meter	% correction (or % error in gas bill)									
	0	0.1	0.2	0.3	0.4	0.5	0.6	0.7	0.8	0.9
	Slow Meters									
0	0.0	0.1	0.2	0.3	0.4	0.5	0.6	0.7	0.8	0.9
1	1.0	1.1	1.2	1.3	1.4	1.5	1.6	1.7	1.8	1.9
2	2.0	2.1	2.2	2.4	2.4	2.6	2.7	2.8	2.9	3.0
3	3.1	3.2	3.3	3.4	3.5	3.6	3.7	3.8	4.0	4.0
4	4.2	4.3	4.4	4.5	4.6	4.7	4.8	4.9	5.0	5.2
5	5.3	5.4	5.5	5.6	5.7	5.8	5.9	6.0	6.2	6.3
6	5.7	6.5	6.6	6.7	6.8	7.0	7.1	7.2	7.3	7.4
7	6.4	7.6	7.8	7.9	8.0	8.1	8.2	8.3	8.4	8.6
8	7.5	8.8	8.9	9.0	9.2	9.3	9.4	9.5	9.6	9.8
9	8.7	10.0	10.1	10.2	10.4	10.5	10.6	10.7	10.9	11.0
10	9.9	11.2	11.4	11.5	11.6	11.7	11.8	12.0	12.1	12.2
11	11.1	12.5	12.6	12.7	12.9	13.0	13.1	13.2	13.4	13.5
12	12.4	13.8	13.9	14.0	14.2	14.3	14.4	14.5	14.7	14.8
13	13.6	15.1	15.2	15.3	15.5	15.6	15.7	15.9	16.0	16.1
14	14.9	16.4	16.6	16.7	16.8	17.0	17.1	17.2	17.4	17.5
15	16.3	17.8	17.9	18.1	18.2	18.3	18.5	18.6	18.8	18.9
16	17.6	19.2	19.3	19.5	19.6	19.8	19.9	20.0	20.2	20.3
17	19.0	20.6	20.8	20.9	21.1	21.2	21.4	21.5	21.6	21.8
18	20.5	22.1	22.2	22.4	22.5	22.7	22.8	23.0	23.2	23.3
19	22.0	23.6	23.8	23.9	24.1	24.2	24.4	24.5	24.7	24.8
20	23.4	25.2	25.3	25.5	25.6	25.8	25.9	26.1	26.3	26.4
21	25.0	26.7	26.9	27.1	27.2	27.4	27.6	27.7	27.9	28.0
22	26.6	28.4	28.5	28.7	28.9	29.0	29.2	29.4	29.5	29.7
23	29.9	30.0	30.2	30.4	30.5	30.7	30.9	31.1	31.2	31.4
24	31.6	31.8	31.9	32.1	32.3	32.4	32.6	32.8	33.0	33.2
25	33.3	33.5	33.7	33.9	34.0	34.2	34.4	34.6	34.8	35.0

Actual reading = (1000 − 40) Mscf = 960 Mscf

To read the meter correction from Table 2.9 for a given meter error (i.e., 4.2 in this case) find the whole percent (i.e., 4) on the left-hand column and find the fractional error (i.e., 0.2) on the top row. Follow respective row and column to find meter correction (i.e., 4 in this case).

2.7   Auditing Tips for Gas Measurement

Auditing gas measurement can be as complex as one can imagine. In particular, with changes in the gas business and application of modern technology, traditional auditing techniques are not adequate. Gas allocation, nomination, measurement, and accounting in a deregulated environment are greatly different from the way they were done in the past. Contractual philosophy and terms have drastically changed in the United States. Multiple contracts at one physical metering point give rise to allocation problems. In addition, time delays in deliveries or redeliveries create accounting imbalances or adjustment problems. Real-time flow information is essential to allocating on the basis of actual volume and to minimizing physical as well as contractual imbalances. In addition, to live by the contractual obligation, the accounting process must take into account the priority of service (firm, interruptible), the allocation method (whether to allocate in proportion to nomination or whether to allocate in whole to firm contracts, with remaining volumes being allocated to interruptible contracts as swing contracts, etc.), overrun, underrun, plant volume reduction (PVR) allocation, etc. Although these accounting processes may seem mind-boggling, auditing physical measurement can be related to one or more physical measurable or unmeasurable factors.

Some of the factors contributing to physical gas imbalance are:

1. *Measurement error:*   Measurement errors are caused by either instrument error or human error. Some common errors are:
   a. Wrong coefficient actual inside diameter (AID) for orifice meter
   b. Wrong plate size
   c. Plate backward
   d. Plate not centered or not down all the way
   e. Meter not in calibration
   f. Incorrect spring range on record
   g. Chart not inked
   h. Painted chart causing square root error
   i. Type of connection (flange/pipe) error
   j. Charts late and closing with estimates
   k. Charts dated wrong
   l. Clock slow or fast
   m. Buckled plate
   n. Chart left in service on a meter run with blind plate
   o. Meter left out of service
   p. Wrong specific gravity applied
2. *Line pack:*   Line pack is the amount of gas contained within the pipeline at anytime that is physically needed to allow the pipeline to operate. Change in line pack can be estimated as shown in Table 2.10 and the example below.
3. *Unknown leaks:*   Unmeasured gas can leak through the distribution company's line, gas lines run by small towns, old valves, etc.
4. *Gas vented by instrument:*   Certain instruments (especially pneumatic controllers) vent gas continually. This may be necessary in order for the instrument to operate.
5. *Blowdown:*   Sometimes incidental amounts of gas are blown

TABLE 2.10 Factors for Computing Change in Line
Pack, Mscf per Mile per Pound Difference

Line size, in	Factor
2	0.0082
4	0.0313
6	0.0711
8	0.1276
10	0.1940
12	0.2900
14	0.3393
16	0.4495
18	0.5752
20	0.7162
22	0.8831
24	1.0441
30	1.6512

down (and not estimated), as necessitated by daily operation. Changing of the orifice plate may require blowing down the upper chamber of a fitting or a meter tube without a dual-chamber fitting. Compliance-type inspection of a regulator may require blowdown. A major blowdown of a section of pipeline for construction or caused by third party damage is usually estimated.

6. *Liquid fallout:* Some hydrocarbons and water vapor may drop out under certain conditions of temperature and pressure. This liquid is equivalent to the gaseous volume that is not accounted for. Water, carbon dioxide, and hydrocarbon knockout at a conditioning, treating, or processing plant should be properly accounted for.

7. Measuring the same gas more than once can contribute to accounting error.

8. Wrong factors or dimensions on record (schedule 40 vs. schedule 80).

9. Wrong cycle on record for a positive displacement or turbine meter [1000 (1M) instead of 10,000 (10M)].

10. Index read and recorded wrong on positive displacement or turbine meter.

Some of the errors listed above can be avoided with careful attention by field operation as well as office personnel. With care and diligence, one can minimize unaccounted-for gas, which is the difference between total input and output.

Example of Use of Table 2.10

43.3 miles of 16-in line

443 lb/in^2 average pressure in line 8:00 A.M., Dec. 24, 1990

396 lb/in^2 average pressure in line 8:00 A.M., Dec. 25, 1990

47 lb/in^2 difference or drop in pressure

$$43.3 \times 47 \times 0.4495 = 915 \text{ Mscf}$$

Since the average pressure at the start of the period was higher, the 915 Mscf is considered as taken from line storage.

2.8 Accounting for Gas Loss

The first step in accurately accounting for gas loss (i.e., accidental blowdown due to rupture) and/or gas used during construction (i.e., drain and purge) is to consistently and accurately measure or calculate the volume of gas lost or used. The second step is proper documentation, including the treatment by accounting to recognize the associated expense. This is necessary to (1) properly charge gas lost or used during construction as a cost of repair or replacement of the pipeline, (2) recover the gas loss expense related to damage of a pipeline by a third party, and (3) to recover the cost of gas used during construction related to relocations or replacements of pipelines to meet state or local government requirements.

Gas loss can be calculated by using the formulas given earlier in "Gas Loss Calculation," Chap. 1. These formulas were derived either from the Weymouth formula or from concepts of critical flow calculation. In any case, these formulas require certain critical information about the actual blowdown, drain, or purge. The biggest gas loss is usually attributable to blowdown, followed by drain loss. In general, purge loss is the smallest part of the total loss, since it is done for a relatively short time at low pressure. The total loss, however, in general consists of blowdown loss plus drain loss plus purge loss.

In addition, to recover expenses related to loss of gas and damages to property by a third party, all pertinent information should be documented. This information includes:

1. A complete description of pipeline or facility damaged.

2. Location of line marker in relation to the damage.

3. Condition of right-of-way (last mowed, trimmed).

4. Was contact made with the party causing the damage? If so, give name and address of person contacted, the date of contact, and the name of the person who contacted the other party.

5. Any leak repair and pipeline replacement report.

To calculate the actual gas loss and to charge these losses appropriately, relevant information should be documented in a form similar to that shown in Fig. 2.8. The accuracy of the calculations depends on the accuracy of information available and assumptions made. In any case, these calculations will provide the best estimates for the information required. Note that, like other flow calculations and measurements, gas loss calculations are not exact and are only as good as the data used. Figure 2.9 shows gas loss per hour through different hole sizes at various pressures.

2.9 Noise

General

Noise is undesired sound. Sound is produced by the transfer of mechanical vibration to air. The air particles in turn produce a variation in normal atmospheric pressure. This variation is characterized by the rate of variation (frequency) and the extent of variation (amplitude). Magnitude of sound intensity affecting the ear varies from 10^{-16} W/cm^2 at the threshold of hearing to 10^{-2} W/cm^2, the region of instantaneous damage.

Mathematically, ear response is shown as

$$R = K \log \frac{I_1}{I_0} \qquad (2.11)$$

where I_0 = base intensity level of 10^{-16} W/cm^2 and I_1 = intensity of a stimulus at a point. R is in bels when $K = 1$.

It has been proven that

$$\frac{I_1}{I_0} \propto \left(\frac{P_1}{P_0}\right)^2 \qquad (2.12)$$

where P_0 is the threshold of hearing at 1000 Hz and equal to a pressure of 0.0002 μbar (*note:* standard atmospheric pressure = 1,013,250 μbar), and P_1 is the pressure at other points away from the source.

Therefore, sound pressure level (SPL) is given by the following relation:

$$\text{SPL} = 10 \log \left(\frac{P_1}{P_0}\right)^2 = 20 \log \left(\frac{P_1}{P_0}\right) \qquad \text{dB} \qquad (2.13)$$

Audible frequencies normally range from approximately 40 to 13,000 Hz, and the range for most noise control work is 100 to 6000

COMPANY
GAS LOSS REPORT

Date of job _____
Name of line _____ Index no. _____
Purpose _____
_____
_____

ACCIDENTAL BLOWDOWN (i.e., rupture)

A. Starting time of blowdown _____ B. Time when valves closed _____
Period of blowdown (B − A) _____
Valves closed: Upstream valve—Mile Pole ____Downstream valve—Mile Pole____
Pressures used for calculations:
    Upstream side of break
$P_1 =$ _____ lb/in^2 gauge $P_2 =$ _____ lb/in^2 gauge
    Downstream side of break
$P_1 =$ _____ lb/in^2 gauge $P_2 =$ _____ lb/in^2 gauge
    (needed if gas was flowing to the hole from both directions)

PLANNED BLOWDOWN OR DRAIN (begins after block valves are closed)

Line size _____ Length of line shutdown _____
Valves closed: Upstream valve—Mile Pole ____ Downstream valve—Mile Pole ____
1. Pressure reduction: from _____   to_____
2. Pressure reduction: from _____   to_____
    (needed if an accidental blowdown occurred and gas was flowing to the hole from
    both directions before block valves closed)

PURGE

Upstream purge pressure _____lb/in^2 gauge
Downstream purge pressure (just upstream of blow-off valve) _____lb/in^2 gauge
Size of blow-off _____ Make and model no. _____ Length of time _____
Specific gravity of purge gas_____
Specific gravity at blow-off at end of purge _____
Blowdown loss _____ Mscf
Drain loss _____ Mscf
Purge loss _____ Mscf
Total loss _____ Mscf
Pipeline leakage section _____ County/parish _____ State_____
Remarks_____
_____
_____

Note: Volumes shall be calculated at a pressure base of 14.73 lb/in^2 absolute. Cal-
         culations and sketch of pipeline section affected must be attached.

Prepared by _____ Date _____ Mscf calculated by _____ Date _____

Figure 2.8   Gas loss report form.

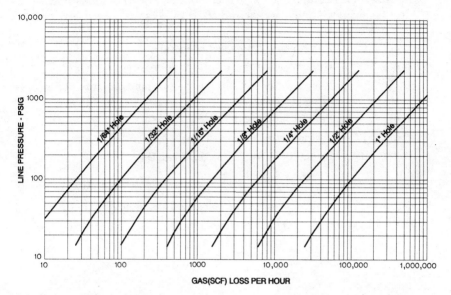

Figure 2.9   Gas loss vs. hole size.

Hz. To control one or more dominating frequency to reduce noise, a frequency survey of noise should be done by frequency-selective sound level meter or spectrum analyzer. Table 2.11 shows permissible noise exposure.

For a *point source of noise* (such as a valve), at a distance x

$$SPL \approx SPL \text{ (source)} - 20 \log \frac{x}{D} \qquad (2.14)$$

For a *line source of noise* (such as aboveground pipeline),

$$SPL \approx SPL \text{ (source)} - 10 \log \frac{x}{D} \qquad (2.15)$$

TABLE 2.11   Permissible Noise Exposure

Duration per day, hours	Sound level, dBA
8	90
6	92
4	95
3	97
2	100
1½	102
1	105
½	110
¼ or less	115

where D = initial distance at which source SPL was measured.
Representative noise levels and their effects are:

175 dB	Can break a thin sheet of metal
160 dB	Human ear drums can be broken
150 dB	Blowdown or relief valve discharging gas into atmosphere; air raid siren
130 dB	Painful to humans
120 dB	Discomfort encountered
70 to 80 dB	Difficulty in telephone conversation
60 to 70 dB	Noise level in offices

Figure 2.10 shows a relation between voice level, distance, and speech interference level.

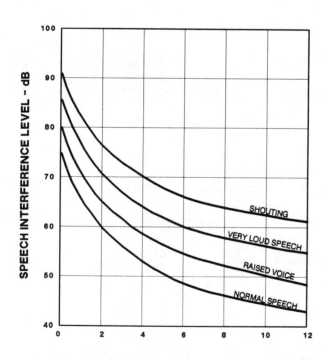

DISTANCE BETWEEN TALKER & LISTENER - ft

Figure 2.10 Voice level and distance vs. speech interference
level. (*Source: Bioastronauts Data Book, NASA SP-3006,
NASA, Washington, D.C., 1964*)

Common noise frequencies are:

Power plant noise	30 to 300 Hz
Fluid dynamic noise (emitted by regulator)	2000 to 10,000 Hz
Speech interference frequency	Three octave bands from 600 to 4800 Hz

Noise prediction for valves and piping

The noise generated by a control valve is caused mainly by three mechanisms: aerodynamic noise, mechanical vibration, and cavitation. Several manufacturers of valves have recently made considerable progress in developing methods for the prediction and reduction of noise. The Fisher Control Co. recommends that the overall sound pressure level in dB be computed as the sum of four variables.

$$L_p = L_p(A) + L_p(B) + L_p(C) + L_p(D) \tag{2.16}$$

where L_p = overall noise level in decibels at a predetermined point
$\quad L_p(A)$ = base sound pressure level in dB, determined as a function of the pressure drop
$\quad L_p(B)$ = correction in dB for gas sizing coefficient
$\quad L_p(C)$ = correction in dB for valve style and pressure drop ratio ($\Delta P/P_1$, P_1 in lb/in^2 absolute)
$\quad L_p(D)$ = correction in dB for acoustical treatment (such as heavy-walled pipe, insulation, in-line silencer)

3

Measurement Engineering
and Design

3.1 General

The basic functions of a measurement facility include the following:

1. Providing a point of gas transfer into or out of company facilities
2. Measuring quantity and quality of gas
3. Controlling pressure or flow of gas stream

The goal is to do this in an accurate, economical, safe, and timely way.

Sound design and engineering of a meter and regulator station requires careful consideration of various aspects. Prior to selection and sizing of primary and secondary devices, standards, etc., one must collect certain engineering information. Final selection and design is also influenced by economic justification under the existing business environment. Therefore, an accurate design with appropriate economic justification must be a goal. In general, in the gas measurement area, the most popular primary devices are the orifice meter, turbine meter, and positive displacement meter. The Annubar device, vortex-shedding flow meter, and insertion turbine meter are other primary metering elements presently available; however, for lack of established industry standards and lack of user acceptability, these are, for all practical purposes, not considered custody transfer quality devices. They are gaining popularity for check measurement or for operational use, primarily for economic reasons.

The secondary devices, traditionally, have been pneumatic and mechanical recorders and controllers. Some of these are (1) an orifice meter with a circular chart that records differential pressure and static pressure, (2) a recording thermometer with a circular chart,

(3) a pressure and volume recorder for a turbine or positive displacement meter that records static pressure as well as cycles based on actual flow on a circular chart. Other mechanical and pneumatic control devices are indicating and/or recording gauges that can control flow or pressure by a final control element (i.e., a control valve) or open or close a valve for meter run switching for better rangeability and accuracy of measurement.

The primary metering and control elements are expected to remain the same for a long time. However, with changing technology and the natural gas business environment, the secondary devices are fast changing. They now not only serve the function of more accurate measurement but also provide faster information service so that timely business decisions can be made. These devices use the established principle of metering, but convert the physical variables to electrical signals instead of chart recordings, which are then used to compute or control flow electronically. In addition, these electronic data acquisition and computing/controlling devices also remotely communicate with one or more host electronic devices that provide remote monitoring, control, and automated billing. Thus, the buyer, seller, and deal maker can mind their shops better by controlling their supply, demand, and inventory.

These electronic devices include electronic transducers, flow computers, remote terminal units (RTU), etc. An electronic transducer converts a physical variable (differential pressure, static pressure, temperature, heating value, specific gravity, etc.) to a standard electrical signal (4 to 20 mA). All of these electronic devices come with various degrees of intelligence ("smartness") and accuracy, and their selection depends on economics and system integration philosophy. More and more, electronic gas measurement and control is becoming an integral part of a broader supervisory control and data acquisition (SCADA) system. Therefore, modern measurement involves several disciplines, including—but not limited to—mechanical engineering, electronics, communications, and computer science. These devices may be powered differently (for example, by solar power, commercial power, or a thermoelectric generator). The communication scheme can involve leased telephone line, microwave communication, very high frequency (VHF) radio, satellite links, cellular phones, etc. Therefore, before getting into the details of selection, design, and engineering, one has to gather relevant information. Some of this information is listed in Table 3.1.

TABLE 3.1 Design and Engineering Data

1. Maximum flow rate _____ MMscf/day

2. Minimum flow rate _____ MMscf/day

3. Normal flow rate _____ MMscf/day

4. Maximum allowable working pressure _____ lb/in^2 gauge

5. Normal measuring pressure _____ lb/in^2 gauge

6. Minimum measuring pressure _____ lb/in^2 gauge

7. Maximum flowing temperature _____ °F

8. Minimum flowing temperature _____ °F

9. Normal flowing temperature _____ °F

10. Base pressure _____ lb/in^2 absolute

11. Base temperature _____ °F

12. Maximum CO_2 (mole %) _____ %

13. Maximum N_2 (mole %) _____ %

14. Specific gravity _____

15. Beta ratio limits _____ to _____

16. Assumed atmospheric pressure _____ lb/in^2 absolute

17. Maximum allowable pressure in interconnecting system _____ lb/in^2 gauge

18. Minimum delivery pressure (downstream of meter) _____ lb/in^2 gauge

19. Maximum allowable noise _____ dB

20. Maximum allowable meter run size _____ inch

21. Gas quality determination method _____

22. Interconnecting piping fluid phase requirement _____

23. Liquid removal requirement _____

24. Liquid reinjection requirement _____

25. Liquid measurement requirement _____

TABLE 3.1 Design and Engineering Data (*Continued*)

26. Location (offshore, onshore, swampland, residential area, etc.) ____

27. Remote monitoring requirement _____

28. Remote control requirement _____

29. Frequency of monitoring required _____
    (real time, daily, weekly, monthly, etc.)

30. Availability of power _____

31. Availability of telephone _____

32. Criticalness of service _____
    (i.e., shutdown of service for inspection allowable or not)

33. Future growth projection (% of flow)_____

34. Odorization requirement _____

35. Condensate removal/storage/handling requirement _____

In addition, the following general specifications should be closely adhered to in building a metering/regulating facility.

Safe practices

The measuring facility will be operated and maintained in a safe manner. Immediate corrective action is necessary to ensure a reliable operating facility and minimize possible property loss and/or personal injury. All local, state, and federal requirements will be followed.

Egress and ingress

Egress and ingress are specifically covered in the contract or agreement. An all-weather road to the measuring facility should be built and maintained year-round with permission from the landowner for its use.

Dehydration facility

Care should be exercised in choosing the type and design of dehydration equipment for different operating conditions. If the dehydration unit is not located immediately upstream of the measuring station, and there is a possibility of liquid accumulation, a free liquid removal device, such as a line drip or separator equipped with a liquid level control valve must be installed directly upstream of the station.

No bypass line should be allowed for passing gas around the dehydration facility into a pipeline.

Compression facilities

When compression facilities are installed it may be necessary to install a pulsation dampener between the compressor and the measuring facility. There are pulsameters available that will indicate the presence of pulsation. Compressor fuel gas should be measured by a separate gas measuring device.

Heating equipment

The installation of heating equipment may be required. This normally occurs when pressure reducing equipment is used and large pressure drops occur. The heating equipment normally consists of a hot water boiler, expansion tank, heat exchanger, circulating pump, three-way modifying valve, and control equipment to maintain a specific temperature. The heating equipment must be located away from the measuring facility.

Flow control and pressure reducing equipment

The capacity of flow control and pressure reducing equipment must meet the minimum and maximum operating conditions of the meter tubes. The equipment will be operated and maintained within the manufacturer's specifications. Isolation valves will be required on each side of the flow control and pressure reducing equipment so the equipment can be maintained and inspected. Any noise level must meet local, state, and federal nighttime and daytime requirements. All gas venting devices will have vent lines piped to the outside of the building.

Over pressure protection

Continuity in grades, ratings, and maximum allowable operating pressure (MAOP) in material is required. Any deviation below MAOP will require over pressure protection.

Meter/regulator station site

A. Location

The station should be located near a public road and/or have an all-weather access road.

All aboveground piping shall be surrounded by a minimum of a 6-foot chain-linked fence with locking gates.

The station area within the fence should be kept weed free through the use of ground cover. This could be plastic covered with gravel, shell, etc.

The site should have adequate drainage to prevent any flooding or standing water.

The site area should be of adequate size for easy access and operation.

The station should be located as close to the pipeline as practical, yet allow sufficient access to the pipeline.

Attention must be paid to local ordinances and regulations governing such facilities (i.e., noise, aesthetics, zoning, etc.).

B. Structures

A walk-in-type building may be required which is suitable for security protection of the measuring equipment and shelter for measurement personnel while calibrating instruments.

The meter building may be mounted on a skid with a metal checker plate floor or installed on a concrete foundation.

Any customer-owned equipment must also reside in a building that is separate from the meter building.

All buildings must have adequate ventilation.

All gas venting devices located inside a building must be piped to the outside.

No flammable material is to be used in the structure or interior of a meter building.

A building with controlled environment shall be erected for chromatograph (control section), printer, and other electronic and communication equipment. It is primarily required for protection of printer paper and to provide shelter for measurement personnel while working on equipment. This building shall be installed in an unclassified area.

3.2 Meter and Regulator Station Design Considerations

Meter station design

AGA Report No. 3 allows for certain minimum and maximum orifice-to-meter tube diameter ratios ($d/D = \beta$) for flange tap and pipe tap connection:

Flange tap $0.15 < \beta < 0.70$

Pipe tap $0.20 < \beta < 0.67$

In general, meter tube sizing is based on $\beta = 0.6$. It is not unusual, however, to fabricate meter tubes for $\beta = 0.75$. This provides for additional capacity in case of dire necessity. Meter tube sizing also depends on meter range (or transducer range if electronic measurement is used). If minimum and maximum flow rates (including future growth potential) can be determined, meter tube size should be based on $\beta = 0.6$ and maximum range of the meter (i.e., 100 or 200 inH_2O). Multiple runs with automatic run switching may be necessary for better rangeability. In addition, a low-range (0 to 10 inH_2O) meter in parallel with a high-range meter may be used on the primary run to measure low flow accurately. Although, for a fixed plate size, an orifice meter has only 3:1 rangeability, in reality the ability to change plates, install multiple runs, stack low-range and high-range meters on the primary run, and use meters with 200 inH_2O or larger range provides very good rangeability.

Turbine meters, on the other hand, have better rangeability, but are essentially fixed for a particular module and do not provide much flexibility. Unlike orifice plates, one cannot change a module size for the run. In general, an orifice meter station is more expensive than a turbine station, but it is highly flexible, fairly accurate, and well-accepted within the industry for custody transfer purposes. Orifice measurement is the most popular method in the United States, particularly among production and transmission companies. Distribution companies, by and large, use positive displacement meters for their low-pressure and low-flow applications. Turbine meters are gaining popularity for certain applications. For space and cost considerations in offshore installations, turbine meters may be recommended for certain applications. The gas industry in Europe is relatively new and overall has fewer meters than the United States. Turbine meters are widely used in Europe.

The specifics of design, selection, sizing, header velocity, etc. are given in "Design and Construction Specifications," Sec. 3.3. Section 3.3 also provides meter run specifications that meet or exceed AGA Report No. 3 recommendations. Figure 3.1 shows a typical dual-orifice meter station.

In general, it is also a common practice in the pipeline industry to use the following chart rotation and test frequencies unless the contract spells them out otherwise:

0 to 10 MMscf/day   Weekly chart/quarterly test

Above 10 MMscf/day   Daily chart/monthly test

Automatic chart changers are used quite often in remote areas or offshore locations, where sending an employee to change charts may not be cost-effective. However, manual chart changing has been found to be far more reliable than automatic chart changing in many instances.

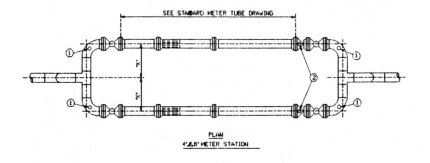

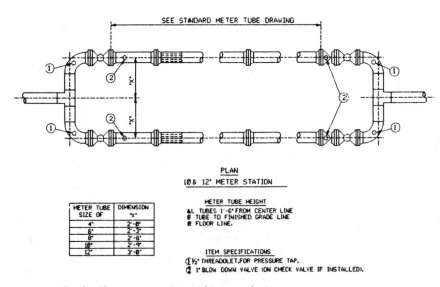

METER TUBE SIZE OF	DIMENSION "X"
4"	2'-0"
6"	2'-3"
8"	2'-6"
10"	2'-9"
12"	3'-0"

METER TUBE HEIGHT
AL TUBES 1'-6" FROM CENTER LINE
B TUBE TO FINISHED GRADE LINE
B FLOOR LINE.

ITEM SPECIFICATIONS
① ½" THREADOLET,FOR PRESSURE TAP.
② 1" BLOW DOWN VALVE (ON CHECK VALVE IF INSTALLED).

Figure 3.1 Dual-orifice meter station without regulation.

With the introduction of electronics and integration of measurement, data acquisition, calculation and control, remote communication and supervisory control, a reliable, accurate, and automated billing mechanism consistent with contractual obligations is expected to be available in the not too distant future.

Regulator sizing

The most popular equation for regulator sizing is the universal gas sizing equation:

$$Q = \sqrt{\frac{520}{G_r T}} \, C_g P_1 \sin \left[\left(\frac{3417}{C_1} \right) \sqrt{\frac{\Delta P}{P_1}} \, \right]^{\circ} \tag{3.1}$$

Under critical (choke) flow condition, this equation reduces to

$$Q = \sqrt{\frac{520}{G_r T}} \, C_g P_1 \tag{3.2}$$

where Q = gas flow rate, scf/h
G_r = gas specific gravity
T = absolute temperature, °R
C_g = gas sizing coefficient
$C_1 = C_g/C_v$
C_v = liquid sizing coefficient; expresses flow rate in gal/min of 60°F water with 1.0 lb/in^2 pressure drop across valve
P_1 = valve inlet pressure, lb/in^2 absolute
$\Delta P = P_1 - P_2$, differential pressure across valve, lb/in^2
P_2 = valve outlet pressure, lb/in^2 absolute

The Becker Precision Equipment (BPE) ball valve regulator sizing equation was developed from the universal gas sizing equation:

$$Q = \sqrt{\frac{520}{G_r T}} \, C_1 C_2 C_v P_1 \sin \left[\left(\frac{3417}{C_1} \right) \sqrt{\frac{\Delta P}{P_1}} \, \right]^{\circ} \tag{3.3}$$

where C_2 = correction factor for variation in specific heat ratio (Table 3.2).

C_1 and C_v values are available from valve manufacturers. In general, high-pressure recovery valves have lower C_1 (around 18) and low-pressure recovery valves have higher C_1 (around 35). Table 3.3 is an example of C_1 (equal to C_A) values for BPE ball valves. In designing a pressure-reducing regulator, there is also a practical limit as to pressure reduction obtainable through a single stage. Figure 3.2 pro-

TABLE 3.2　Values of C_2 and G_r for Various Gases

Gas	C_2	G_r
Acetylene	0.98	0.90
Air	1.00	1.00
Butane	0.94	2.00
Ethane	0.96	1.03
Helium	1.04	0.14
Hydrogen	1.00	0.07
Methane	0.98	0.55
Natural gas (representative)	0.98	0.60
Nitrogen	1.00	0.97
Oxygen	1.00	1.10
Propane	0.95	1.52
Propylene	0.91	1.45

TABLE 3.3　C_1 Values for BPE Ball Valves

BPE ball valve bore interior diameter (ID), in	Adjacent line ID, in	C_A
1	1	14
1	1½	16
1	2	20
1	2½	22
1	3	25

vides a way to determine number of stages of regulation required for high-pressure applications.

Station design considerations

As evident from Table 3.3, C_1 values are influenced by connected piping. In the past, pressure losses through piping, fittings, and valves were not given much consideration in station design. With high-capacity ball valve regulators, however, the station piping configuration becomes very important.

To minimize problems with noise and vibration and possible resulting damage, it has become a common practice to design the regulator station piping so that the velocity of the gas through the station is limited to 200 ft/s for buried valves and 100 ft/s for aboveground valves.

The velocities through regulator station piping can be calculated by the following formula:

$$V = \frac{0.75Q_h}{Pd^2} \tag{3.4}$$

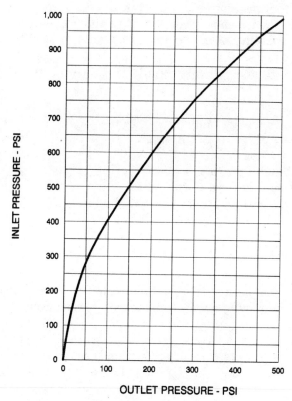

Figure 3.2 Practical pressure reduction and regulator stages required.

where V = velocity, ft/s

Q_h = flow rate, scf/h

P = inlet pressure, lb/in^2 absolute

d = inside diameter of the station piping, in

Rearranging the above formula gives an equation for the station piping size:

$$d = 0.866 \sqrt{\frac{Q_h}{PV}} \tag{3.5}$$

Another feature of the regulator station to be considered is the *turndown ratio* of the ball valve regulator, which is the ratio of the maximum inlet pressure to the desired outlet pressure of the ball valve regulator. The ball valve, being a double-seated valve, will function at a wider angle of opening than will a single-seated valve. This is a result of the double pressure reduction through the valve itself. The first reduction occurs across the upstream seat, and the second, across the downstream seat. This phenomenon gives the ball valve regulator a high turndown ratio capability. A maximum 100:1 ratio is used by some as design practice.

The turndown ratio TDR of a ball valve regulator can be calculated by the following formula:

$$\text{TDR} = \frac{\max Q_h}{\min Q_h} \sqrt{\frac{\max P_1}{\min P_1}} \sqrt{\frac{\max \Delta P}{\min \Delta P}} \qquad (3.6)$$

where $\max Q_h$ = maximum flow rate, scf/h
$\min Q_h$ = minimum flow rate, scf/h
$\max P_1$ = maximum inlet pressure, lb/in^2 absolute
$\min P_1$ = minimum inlet pressure, lb/in^2 absolute
$\max \Delta P$ = maximum differential pressure, lb/in^2 absolute
$\min \Delta P$ = minimum differential pressure, lb/in^2 absolute

Figure 3.3 shows an example of a multiple-tube orifice meter station with single- or two-stage regulation. If single-stage pressure control (reduction) is desired, normally the regulator station should be upstream of the meter to provide a steady metering pressure. Gas temperature is expected to drop with pressure reduction through a regulator. A rule of thumb estimate is a 7°F drop for each 100 lb/in^2 reduction in pressure. Therefore, special care must be taken to avoid possible freezing conditions, especially with moisture or any condensate vapor in the gas stream. Figure 3.2 can be used to determine the number of regulators required in series of given pressure conditions as well as to determine practical pressure reduction.

Example   Practical pressure reduction and stages of regulation required for high-pressure applications can be determined by the following procedure. (Refer to Fig. 3.2.) Given: inlet pressure = 700 lb/in^2; desired outlet pressure = 20 lb/in^2.

Step 1:   Locate 700 lb/in^2 gauge on inlet scale. Trace horizontally to the right of the curve. Then trace downward to the outlet pressure scale. Obtain outlet pressure of 250 lb/in^2 gauge.

Step 2:   Follow Step 1, starting with inlet pressure of 250 lb/in^2 gauge. Obtain outlet pressure of 50 lb/in^2 gauge.

Step 3:   Follow Step 1 with inlet pressure of 50 lb/in^2 gauge. Outlet pressure is found to be less than the desired outlet pressure of 20 lb/in^2 gauge. Therefore, this regulator will be set at the desired outlet pressure.

Therefore, three stages of pressure reduction are required: 700 to 250 lb/in^2 gauge, 250 to 50 lb/in^2 gauge, and 50 to 20 lb/in^2 gauge.

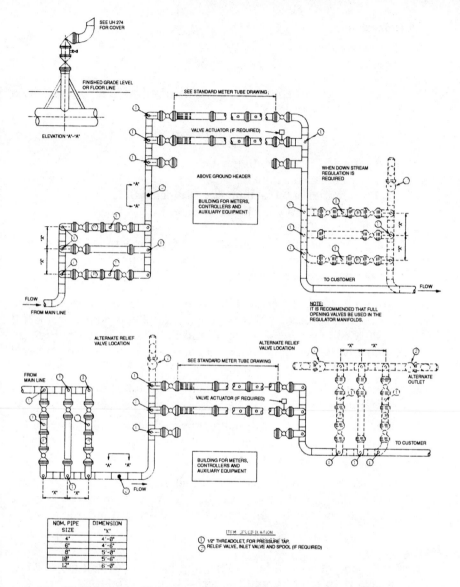

Figure 3.3   Multiple-tube orifice meter station with one- or two-stage regulation.

3.3   Design and Construction Specifications

The following specifications are provided as a general set of guide-lines. Exact specifications may vary from company to company.

3.3.1 Mechanical

A. General piping rules
Design factor

1. The design pressure for steel pipe is determined in accordance with the following formula:

$$P = (2St/D)FET \tag{3.7}$$

where P = design pressure, lb/in^2 gauge
S = yield strength, lb/in^2
D = nominal outside diameter of pipe, in
t = nominal wall thickness of pipe, in
F = design factor
E = longitudinal joint factor as given for various pipe classes by U.S. Department of Transportation (DOT)
T = temperature derating factor

A design factor F of 0.50 shall be used in the design formula, except that meters located within refinery limits are subject to ANSI B31.3, "Code for Chemical Plant and Petroleum Refinery Piping." A temperature derating factor of 1 shall be used for temperatures of 250°F or less.

2. The entire meter station facility shall be designed to a minimum of the mainline pressure whether regulation is used or not.

3. Meter run header velocity must not exceed 50 ft/s. Regulator piping velocity should not exceed 100 ft/s aboveground and 200 ft/s belowground. Refer to pipeline velocity chart (Fig. 3.4)

Testing. Station piping shall be tested in accordance with established procedures of the company, but must meet or exceed state or federal requirements as well as accepted industry standards.

Valves

1. Meter run valves will be full-opening ball valves and may be one size less than the ultimate run size.

2. Bypass valves will be full-opening soft-seat block and bleed-type gate or ball valves and are the same size as the bypass.

3. Blow-off valves shall be 1-in ball valves installed on each end of the meter run.

4. Check valves will normally be installed to avoid possible backflow through the meter run, unless the meter station is designed for bi-directional flow.

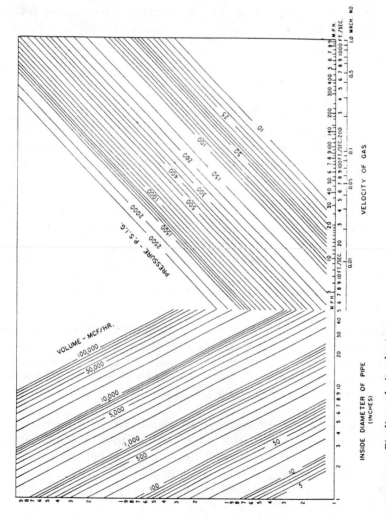

Figure 3.4 Pipeline velocity chart.

Headers

1. Headers for multiple-run meter installations shall be designed to give as even a distribution of flow as possible. Normally the header shall have a K value (Weymouth formula) equal to the summation of the K values of the meter tubes connected to the header. The K value shall be based on the specific gravity of the gas being measured, where K stands for the flow factor of the meter tube or header as appropriate ("Orifice Meter," Sec. 3.3.1B)
2. Headers shall have drawn nozzle outlets and/or reducing welding tees and elbows.

Fittings

1. When reducers and reducing elbows are used, they shall be installed such that there shall be no liquid trap in the meter tube.
2. All flanges and valves shall be of appropriate ANSI rating (i.e., ANSI 150, 300, 600, or 900) and must not lower the maximum allowable operating pressure (MAOP) of the overall system.

Gauge lines

1. All gauge lines longer than 20 ft shall be 1-in size to the interior of the building. ½-in pipe or ⅜- × 0.035-in WT (wall thickness) 304 stainless-steel tubing shall be used for gauge lines shorter than 20 ft. The gauge lines shall be sloped back to the meter run to avoid liquid traps. A minimum slope of ⅛ in for each linear foot should be maintained.
2. Frost heaving and loose backfill settling around buried small piping and tubing can cause severe leaking problems. Sensing lines to regulators, telemetering pressure lines, and gas power and control lines to valve operators shall generally be above grade or properly supported to prevent breakage.

Meter houses. Meter houses shall be walk-in type. The style of house shall be dictated by circumstances. 4- × 6-ft and 6- × 8-ft houses are normally galvanized. For larger than 6- × 8-ft, steel side panels and white enamel roof panels are commonly used.

Where sour gas is involved, small non-walk-in enclosures will be used to house the meter instrumentation.

Supports. Horizontal supports for meter runs shall be made of pipe to prevent corrosion.

B. Orifice meter

Meter tube and header size. Diameter of meter tubes shall be based on a maximum β ratio of 0.60 and maximum operating differential range (i.e., 100 or 200 in) of meter or transducer. Headers and risers shall be sized and spaced for the ultimate estimated load. Headers are sized according to the Weymouth formula:

$$M = (ID)^{8/3} \times \frac{15,387}{\sqrt{520 G_r}} \qquad \text{at 14.65 lb/in}^2 \text{ pressure base} \qquad (3.8)$$

where M = flow factor for meter tube and G_r = 0.60 (typical). The header flow factor is

$$K = M \times \text{number of runs} \qquad (3.9)$$

Calculate the M value for single run in by Eq. (3.8). Determine the K value from Eq. (3.9), then substitute this value of K for M in Eq. (3.8) and solve for header ID.

In general, a straight run with no header is satisfactory for a single meter run. For balanced flow, a T-type header (of the same size as a meter tube) for a two-run station and a U-type header for more than a two-run station are recommended.

Header velocity should not exceed 50 ft/s. For a U-type header, header size can be calculated by using K and M values, as shown earlier. A much simpler way to size the header is to make its cross-sectional area equal to the sum of all the meter tube cross-sectional areas. These methods will assure maintaining header velocity below 50 ft/s. See Table 3.4 for header size for multiple meter runs.

TABLE 3.4 Header Size for Multiple Meter Runs, Inches

Number of tubes	Tube size								
	3	4	6	8	10	12	16	20	24
2	6	6	10	12	16	18* (20)	24	30	34* (36)
3	6	8	12	16	18* (20)	22* (24)	28* (30)	36	42
4				16	20	24	32* (36)		
5					24	28* (30)	36		
6					26* (30)	30	42		
7						32* (36)			

*Calculated size, but not normally used. Use larger size shown in parentheses.

Insulating flanges. Insulation shall be provided at change of ownership. Any tubing, piping, or conduit that may be bypassing the insulating flanges shall be installed with insulated unions or bushings. Care shall be taken to prevent thermal expansion or contraction that could cause excessive stresses.

Drips. Filter-separators or "drips" shall be installed upstream of all meter runs measuring undehydrated gas or gas that has not been processed through a gasoline plant. Drips in DOT Class 1, 2, or 3 locations shall have a design factor of 0.50. Drips in Class 4 shall have a design factor of 0.40.

Orifice meters

1. Meter runs shall meet or exceed the requirements of AGA Gas Measurement Committee Report No. 3, latest revision. See capacity information on sizing (Table 2.1). See Figs. 3.5 and 3.6 for dimensional requirements that meet or exceed AGA Report No. 3 (1985 edition).

2. Meter runs shall be made of commercial meter tubing. Carefully selected smooth steel pipe may be substituted for commercial meter tubing when material availability does not meet construction schedules. Inside pipe walls shall be smooth, with roughness not to exceed 300 μin.

3. Meter runs will normally be equipped with "senior-type" (double-chambered) orifice fittings, FE (flanged end), WE (weld end).

4. "Junior-" or standard-type orifice fittings, WE, FE, will be considered as special cases.

5. All meter runs shall have flanges on both upstream and downstream ends. Meter runs shall have a ½-in spacer installed in the downstream flange. Jackscrews shall be installed on the flanges at the spacer location. Roll bars shall be provided to allow inspection for foreign material and centering of the orifice plate.

6. Ring-type straightening vanes shall be installed upstream as recommended in AGA Report No. 3. On bidirectional tubes, such straightening vanes are required on both sides. Manufacturers' standard-length vanes meet AGA Report No. 3.

7. See Fig. 3.7 for required taps on meter tubes.

8. Flange taps shall be used on all orifice meters. Dual-tap orifice fittings shall be installed to satisfy immediate or future needs for telemetering or check measurement by others.

9. See Fig. 3.8 for details of orifice plates.

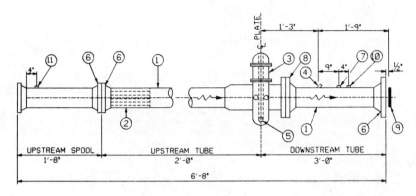

ITEM SPECIFICATIONS

① GRADE B PIPE OR TUBING OF METER TUBE TOLERANCE,PER AGA*3(0.5% TOLERANCE)WITH 0.75
   BETA,(2.067" I.D.,.154' WALL,3.65"/FT.,STD.WALL)OR(1.939"I.D.,.218" WALL,
   5.02"/FT.,XH WALL).

② TUBULAR STRAIGHTENING VANE FLANGE TYPE 5.0 INCHES LONG.

③ ORIFICE FITTING FE x WE DOUBLE CHAMBERED C/W 2-SETS OF FLANGE TAPS
   PER AGA*3 WITH 0.75 BETA,   PSIG WP (0.5 DF),ANSI

④ 45° LATERAL CONNECTION,DRILLED FOR 1' SEPARABLE SOCKET.
   (FOR THERMOMETER BULB OR R.T.D.)

⑤ ½' OR ¾' DRAIN.

⑥ WELD NECK, RAISED FACE FLANGE ANSI   (I.D. TO MATCH I.D. OF PIPE).

⑦ ½' XH THREADOLET WITH ½' NPT TEST WELL, C/W CHAIN & PLUG.

⑧ ON ANSI 400 AND HEAVIER-RECESSED FACE,WELD NECK FLANGE,ANSI   MACHINED TO INSURE
   GOOD ALIGNMENT WITH FITTING.
   ON ANSI 300 AND LIGHTER-RAISED FACE,WELD NECK FLANGE,ANSI   C/W CLOSE TOLERANCE
   DOWELL PINS (TWO MIN.) INSTALLED TO INSURE GOOD ALIGNMENT WITH FITTINGS.

⑨ ½' THICK CARBON STEEL SPACER PLATE,I.D. TO MATCH TUBE BORE. O.D.TO FIT INSIDE
   BOLT CIRCLE.

⑩ ½' XH THREADOLET WITH SAMPLE PROBE (1/2 'M x ¼'F PROBE).

⑪ ½' XH THREADOLET FOR BLOW-DOWN VALVE.

NOTES:

1. ALL TAPS TO BE INSTALLED ON TOP OF PIPE.
2. LENGTH OF SEPARABLE SOCKET (i.e.,THERMOWELL) TO BE 3', EXCLUDING THREADS.

Figure 3.5   2-in ANSI meter tube with senior fitting (adapted for flange taps only).

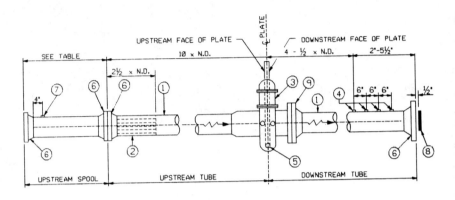

NOTE:
N.D.= NOMINAL DIAMETER (i.e. 8")
A.I.D.= ACTUAL INSIDE DIAMETER (i.e. 7.981")
ALL TAPS TO BE INSTALLED ON TOP OF PIPE

ITEM SPECIFICATIONS

① PIPE OR TUBING OF METER TUBE TOLERANCE , AS PER AGA # 3, (0.5% TOLERANCE)
WITH 0.75 BETA, 'I.D., "WALL   "/FT.,SMYS=   PSI, W.P.(0.5 DF)=   PSIG.

② TUBULAR STRAIGHTENING VANE (4'& 6'-FLANGE TYPE; 8'THRU 16'-FLANGE AND PIN LOCK),
PER AGA # 3.

③ ORIFICE FITTING FE × WE DOUBLE CHAMBERED C/W 2-SETS OF FLANGE TAPS
PER AGA # 3 WITH 0.75 BETA,   PSIG W.P. ANSI

④ 1' THREADOLET, DRILLED FOR 1' SEPARABLE SOCKET.

⑤ ½' OR ¾" DRAIN.

⑥ WELD NECK, RAISED FACE FLANGE ANSI   (I.D. TO MATCH I.D. OF PIPE).

⑦ 1' THREADOLET FOR BLOW-DOWN VALVE.

⑧ ½' THICK CARBON STEEL SPACER PLATE, I.D. TO MATCH TUBE BORE. O.D. TO FIT INSIDE BOLT CIRCLE.

⑨ ON ANSI 400 AND HEAVIER-RECESSED FACE, WELD NECK FLANGE, ANSI   MACHINED TO INSURE
GOOD ALIGNMENT WITH FITTING.

ON ANSI 300 AND LIGHTER-RAISED FACE, WELD NECK FLANGE, ANSI   C/W CLOSE TOLERANCE
DOWELL PINS (TWO MIN.) INSTALLED TO INSURE GOOD ALIGNMENT WITH FITTINGS.

DIMENSIONS

TUBE SIZE	UPSTREAM SPOOL	UPSTREAM TUBE	DOWNSTREAM TUBE (WITH SPACER)	TOTAL LENGTH	VANE LENGTH
4'	2'-8'	3'-4'	4'-0'	10'-0'	0'-10'
6'	4'-7"	5'-0'	4'-9"	14'-4'	1'-3'
8'	6'-6'	6'-8'	5'-6"	18'-8'	1'-8'
10'	8'-5'	8'-4'	6'-3'	23'-0'	2'-1'
12'	10'-4'	10'-0'	7'-0'	27'-4'	2'-6'
16'	14'-2'	13'-4'	8'-6'	36'-0'	3'-4'

Figure 3.6   Orifice meter tubes with orifice fitting (adapted for flange taps only).

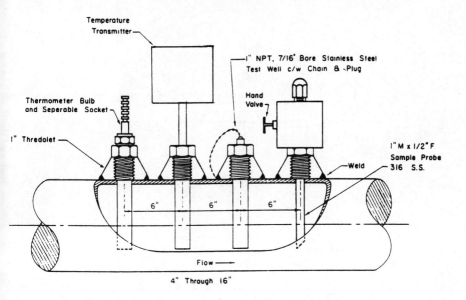

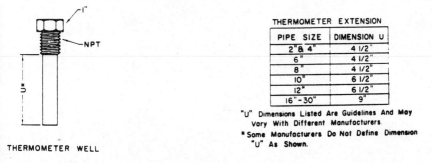

Figure 3.7   Typical thermometer and test well installation.

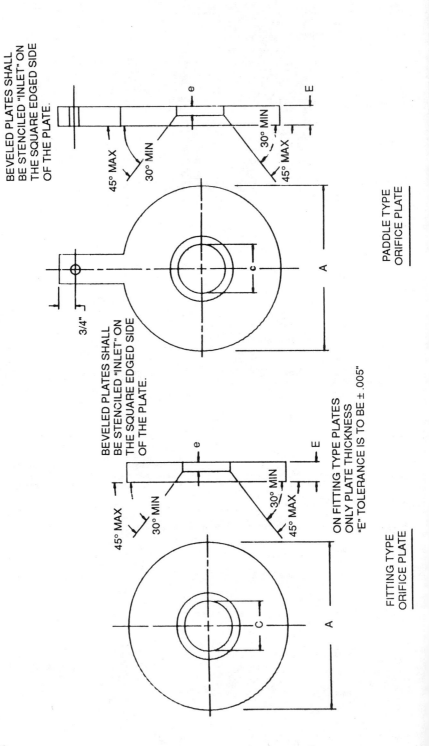

BEVELED PLATES SHALL
BE STENCILED "INLET" ON
THE SQUARE EDGED SIDE
OF THE PLATE.

45° MAX
30° MIN
30° MIN
45° MAX

e

E

A

C

PADDLE TYPE
ORIFICE PLATE

3/4"

BEVELED PLATES SHALL
BE STENCILED "INLET" ON
THE SQUARE EDGED SIDE
OF THE PLATE.

45° MAX
30° MIN
30° MIN
45° MAX

e

E

ON FITTING TYPE PLATES
ONLY PLATE THICKNESS
"E" TOLERANCE IS TO BE ±.005"

A

C

FITTING TYPE
ORIFICE PLATE

92

The Dimension Specifications Of The Paddle Type Plates Allow The Use Of These Plates in Both Of The ANSI 300, Std. Orifice Flanges, and the ANSI 600 Orifice Flange With The Proper Spacer Rings For Centering.

For Bi-Directional Stations, Order Double Edge Plates Only. Plates To Be Made Of 304 Stainless Steel.

Note:
The Upstream Face Of Orifice Plates Shall Be As Flat As Can Be Obtained Commercially. This Flatness Must Not Depart From True Along Any Diameter By More Than 0.01' Per Inch Of The Dam Height (A–d)/2. The Upstream Face Of The Orifice Plates Shall Have A Finish At Least Equivalent To That Obtained In Commercial Cold–Finished Sheet –Stock Of Alloy Stainless Steel With No. 4 Polish Or Better.

TOLERANCE FOR ORIFICE DIA.

ORIFICE SIZE "d"	TOLERANCE PLUS OR MINUS
0.250"	0.0003"
0.375"	0.0004"
0.500"	0.0005"
0.625"	0.0005"
0.750"	0.0005"
0.875"	0.0005"
1.000"	0.0005"
OVER 1.000"	0.0005" PER INCH d

DIMENSIONS IN INCHES

Pipe Size	Dimension A		Paddle	B	C	E	e
	Daniels	Peco					
2	2.437	2.437	4-3/8	4	1-1/2	1/8	1/32
3	3.437	3.437	5-7/8	4	1-1/2	1/8	1/32
4	4.406	4.406	7-1/8	4	1-1/2	1/8	1/16
6	6.437	6.437	9-7/8	4	1-1/2	1/8	1/16
8	8.437	8.437	12-1/8	6	1-1/2	1/8	1/8
10	10.687	10.687	14-1/4	6	1-1/2	1/4	1/8
12	13.079	12.593	16-5/8	6	1-1/2	1/4	1/8
16	16.583	16.000	21-1/4	6	1-1/2	3/8	1/8
18	18.563	17.938	23-3/8	6	1-1/2	3/8	1/8
20	20.583	20.000	25-5/8	6	1-1/2	3/8	3/8
24	24.500	24.250	30-3/8	6	1-1/2	3/8	3/8
30	30.750	30.500	37-3/8	6	1-1/2	1/2	1/2
36	36" Plates Should Be Ordered Using Serial No. Of Existing Plates.						

Figure 3.8 Orifice plates.

Meter tube specification, inspection, and calibration. Figure 3.9 provides meter tube specifications and inspection, calibration, and testing procedures, with necessary forms. These procedures are extremely important to ensure accuracy of measurement.

C. Turbine, positive displacement, and other meters

Turbine meter

1. Meter runs shall meet or exceed the requirements of AGA Gas Measurement Committee Report No. 7, latest revision. Refer to "General piping rules" in Sec. 3.3.1 for additional design guidelines and requirements.

2. The companion flanges to the turbine meter shall have the internal weld bead ground flush to eliminate abrupt variations in the internal diameter of the pipe immediately adjacent to the meter.

3. For dimensional requirements, vane location, strainer location, tap location, etc. refer to Figs. 3.10 and 3.11.

4. An in-line straightening vane shall be installed upstream of the turbine meter so that there will be a minimum distance of 5 pipe diameters between the outlet end of the vane and the meter inlet.

5. A minimum of 2 pipe diameters without attachments shall be provided downstream of the meter. Then a welded tee reducing outlet shall be installed for pressure monitoring and blowdown. The blowdowns shall be sized as follows:

Meter run size, in	Blowdown coupling size, in
2	¼
3	½
4	½
6	1
8	1
12	1

6. This blowdown tap can be used for reasonability test with a sonic nozzle prover. Flow rate should be maintained above 10 percent of the meter capacity. Typically when flow rate falls below 3 percent of capacity, a turbine meter should be resized. A routine spin test with acceptable spin time recommended by the manufacturer assures satisfactory performance. However, a turbine module can be high-pressure-tested and proved in a standard body within an accuracy of ±0.5 percent as suggested by manufacturer.

Figure 3.9 Meter tube specification, inspection, and calibration.

(a) METER TUBE SPECIFICATIONS

1. Sizes and dimensions to be shown on drawings. Overall length of upstream or downstream tube not to exceed ± ⅛ in from drawing.
2. All fabrications and tolerances to conform with AGA Gas Measurement Committee Report No. 3 for 0.75 β ratio, out-of-round tolerances is to be 0.2 percent × AID for tubes with pipe taps and flange taps, 0.5 percent × AID for tubes with flange taps only. Measured AID does not have to agree with published AID coefficients. The AID coefficients will be calculated from AID.
3. Locations of the pipe and flange taps to conform to AGA Report No. 3 specifications. Tap hole variation not to exceed allowable tolerance for 0.75 β ratio. It is to be noted that use of pipe tap is discouraged, since flange tap measurement is most commonly accepted within the industry.
4. Inside of all sections of meter tube shall be thoroughly cleaned to remove all rust, scale, etc. Inside of tube and fitting shall be thoroughly coated with motor oil. Working parts of fitting shall be thoroughly greased to ensure smooth operation and protection from corrosion.
5. All pipe utilized in the fabrication of meter tubes and all welding shall meet the requirements of the latest edition of the Department of Transportation, Title 49, Part 192, Code of Federal Regulations. Mill test reports shall be provided.
6. The upstream and downstream tubes of all meter tubes will be bored and/or honed as required to comply to AGA Report No. 3 requirements for inside pipe roughness and defects. Specified wall thickness is the finished wall thickness after any boring or honing operation.
7. All fabrications shall be sandblasted and cleaned to be in compliance with near-white metal abrasive blast cleaning standards established by National Association of Corrosion Engineers (NACE).
8. All fabrications are to be shipped to owner with an application of inorganic zinc primer coat. Intermediate and/or final coating not required unless specified otherwise. Primer shall be applied *after* the hydrostatic test and according to painting manufacturer's recommendations sheet.
9. All flange stud bolts, American Society for Testing and Materials (ASTM) A-193, Grade B-7, and hex nuts, ASTM A-194, Class 2-H, shall be cadmium-plated.
10. Maximum allowable flange tilt to be ¼°. Flange faces to be serrated.
11. Additional requirements: _____ .

ORIFICE FITTING SPECIFICATIONS

1. Orifice fitting to be _____ WP (Working Pressure), ANSI, RF (Raised Face) flanged on downstream side and to be beveled for welding on upstream side, unless otherwise noted on drawing _____ .
2. Orifice fitting to be bored to match ID of meter tubes.
3. Carrier and plate to be of double-sealed unit type or equal for two-way flow.
4. Orifice fitting is to be complete with two sets of flange taps as per AGA Report No. 3 with allowable tolerance for 0.75 β ratio.

NOTIFICATION REQUIREMENTS

1. All circumferential welds are to be x-ray–inspected and filmed, and certification of inspection is to be forwarded to _____ .
   Attention: _____ .

(Continued)

Figure 3.9 (*Continued*)

2. All tubes and fittings are to be tested hydrostatically to _____ lb/in² (1.5 × WP) for 8 h prior to calibration. After 8-h test, where double-chambered fitting is used, the isolating valve between chambers shall be tested 1 h at a differential of _____ lb/in² (1 × WP). No welding shall be allowed after hydrostatic test. A copy of recorded test chart showing both tests shall be forwarded to ___ _____ . Attention: _____ .

3. _____ days prior to shipping meter tubes, and after successful completion of hydrostatic test and blank plate leakage test, the Superintendent of Measurement is to be notified for field inspection.

4. Do not ship until notified by Purchasing Department.

5. Mail tube calibration sheet (signed by witness) to _____
_____ .

(*b*)   METER TUBE INSPECTION CHECK SHEET

I. Tube number _____ Date _____

   Fitting number _____ Plate holder type _____

   Manufacturer _____ Size _____ ANSI _____

   Destination of runs _____ Purchase Order number _____

   Inspector _____ Authorization for Expenditure number _____

	Good	No good
II. Welding		
A. Seams of flanges	☐	☐
B. Seams of tube turns	☐	☐
C. Weld on taps and wells	☐	☐
III. Measurements	Good	No good
A. Tap holes right size	☐	☐
B. Tap hole location (2½ by 8D for pipe taps, 1 in for flange taps)	☐	☐
C. Straightening vane location	☐	☐
D. Length of pipe all right	☐	☐
E. Out-of-roundness, upstream tube (mike), shall not exceed 0.2 percent of coefficient AID for pipe taps and 0.5 percent for flange taps	☐	☐
F. Out-of-roundness, downstream tube (mike), shall not exceed 0.4 percent of coefficient AID for pipe taps and 1.0 percent for flange taps	☐	☐
G. Plate off center, less than 3 percent	☐	☐
IV. Miscellaneous	Good	No good
A. Pipe free of grooves, pits, or deep scratches	☐	☐
B. Flanges mounted straight on pipe	☐	☐
C. Tap holes slanted	☐	☐
D. Tap hole leak test	☐	☐

V. Buyer/vendor drawing checked to meter tube specs   Yes ___   No ___

VI. Hydrostatic test chart attached   Yes ___   No ___

VII. Vendor calibration sheets attached   Yes ___   No ___

REMARKS: _____

(c) PROCEDURE FOR BLANK PLATE LEAKAGE TEST

1. Isolate and vent the orifice meter run. Ensure that the upstream and downstream isolating valves do not leak.
2. Install an unbored orifice plate in the orifice fitting.
3. Attach the test manifold as shown in the recommended test manifold illustration.

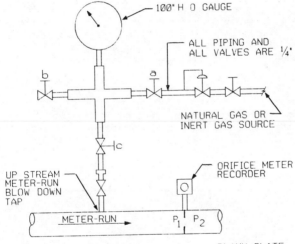

4. Connect the pressure source to the inlet to the manifold regulator.
5. With valve b partly open, valve a fully open, and valve c closed, adjust the regulator to deliver 55 inH$_2$O (approximately 2 lb/in^2 gauge).
6. Close valve b and open valve c and any other valves required to admit gas from the manifold to the orifice meter run.
7. When the orifice meter run is pressured up to 55 inH$_2$O, close valves a and c and open valve b. If no leakage is observed at valve b, proceed with the test.
8. Observe the orifice meter recorder for 10 minutes. If no differential pressure decrease is observed, proceed to the next step. If a decrease is observed, the orifice meter run must be disassembled and inspected by using liquid leak detector or some other method to check seal and tap hole integrity.

(*Continued*)

Figure 3.9   (*Continued*)

9. Vent enough gas through valve *c* to reduce the pressure to 10 inH$_2$O. Observe the orifice meter recorder for 5 minutes. If no differential pressure decrease is observed, proceed to the next step. If a decrease is observed, follow the steps detailed in Step 8.
10. Return the orifice meter run to normal operation.
11. Attach orifice meter chart to test report.

ORIFICE FITTING BLANK PLATE LEAKAGE TEST REPORT

Vendor: _____   Date _____

Location of test _____

Orifice fitting serial no. _____

Nominal size _____

ANSI class _____

Meter tube serial no. _____

Test method (attach procedure or describe)

_____

_____

_____

Orifice meter recorder calibration date _____

Test gauge calibration date _____

Results: _____

_____

_____

_____

Inspector's signature _____   Date _____

Vendor's signature _____   Date _____

(d) ORIFICE METER TUBE CALIBRATION REPORT

To
Measuring Gas From _____ Station No. _____

Station Location _____ Pur. Ord. No. _____

Fabricated By _____ Tube Ser. No. _____

Orifice Fitting Make _____ Type _____ Ser. No. _____

Straightening Vanes: Yes _____ No _____   Type Vanes: Pinned _____ Flanged _____

Meter Run Type No. _____   Type Flanges (S.O./W.M.) _____   A.N.S.I. _____

TYPE NO. 1

TYPE NO. 2

TYPE NO. 3

*Note: Cross-section 'A' and 'F' shall be at two pipe diameters except if such cross-section falls within the fitting or at weld it shall be made 1" inside the pipe.

Type No. 3 and other meter tubes set up for reverse flow: Stencil individual A.I.D. on each tube section and fitting.

Per AGA 3, Second Edition, 1985
Coefficient A.I.D. is determined from measurements taken in a plane 1 inch upstream from the upstream face of the orifice plate.

Designate Upstream or Downstream tube in section below.

Type No. _____	Stream Tube				Stream Tube			
	A	B	C	G	D	E	F	H
V								
LV								
RV								
H								
Average = C or G and D or H Stamped on Tube								
Max. Dia - Min. Dia.		=			—	=		

Coefficient A.I.D. (Normal flow) = _____   Coefficient A.I.D. (Reverse flow) = _____

Meter Tap Distance From Each face of Plate. { Upstream _____ Inches (P.T.)   _____ Inches (P.T.)
_____ Inches (F.T.) Lt.   Downstream   _____ Inches (F.T.) Lt.
_____ Inches (F.T.) Rt.   _____ Inches (F.T.) Rt.

Remarks _____

Inspected By: _____ Checked By: _____

For: _____ For: _____

Date: _____ Date: _____

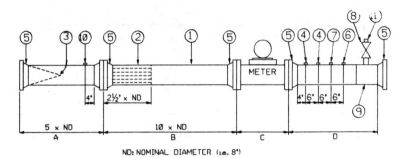

ND: NOMINAL DIAMETER (i.e. 8")

ITEM SPECIFICATIONS

① GRADE B SEAMLESS PIPE   "I.D., "WALL, *⁄FT.,W.P.(0.5 D.F.)   psig.
② TUBULAR STRAIGHTENING VANE (4"& 6" FLANGE TYPE; 8"-12" FLANGE & PIN LOCK)
③ BASKET STRAINER: 150% STRAINER OPENING WITH   x   S.S.WIRE MESH.
④ 1" THREADOLET,DRILLED FOR 1" SEPARABLE SOCKET. ONE FOR THERMOMETER BULB,OTHER
   FOR R.T.D. AS APPLICABLE
⑤ WELDING NECK FLANGE,RAISED FACE,ANSI
⑥ 1" THREADOLET FOR BLOW-DOWN.(FOR 4" TUBES USE½" VALVE. FOR 6",8",AND
   12" TUBES USE 1" VALVE.)
⑦ 1" THREADOLET, 1" NPT⁷⁄₁₆" BORE,STAINLESS STEEL TEST WELL C/W CHAIN & PLUG.
⑧ 2" F.O.,S.F.,BALL VALVE,ANSI
⑨ REDUCING OUTLET,WELD TEE: SIZE   ,STD. OR X-HEAVY,(4"x 2" WELD REDUCER
   ALSO REQUIRED FOR 8" & 12" TUBES UNLESS EXTRUDED. REDUCING OUTLET WELD
   TEE IS AVAILABLE).
⑩ 1" THREADOLET WITH SAMPLE PROBE. (1" M x ½" F PROBE)
⑪ 2" BULL PLUG

SIZE	ANSI	A	B	C	D
4"	300 600	1'-8"	3'-4"		3'-3"
6"	300 600	2'-6"	5'-0"	SEE MANUFACTURER'S DRAWINGS	3'-8"
8"	300 600	3'-4"	6'-8"		5'-0"
12"	300 600	5'-0"	10'-0"		6'-11"

NOTES:
1. INSTALL 1" THREADOLET JUST U.S.
   OF U.S. ISOLATING VALVE.
2. FOR ANGLE BODY METER USE SAME
   U.S. AND D.S. SPOOLS AS SHOWN
   ABOVE
3. PIPE INTERIOR IS TO BE OF COMMERCIAL
   ROUGHNESS,AND THE FLANGE I.D.'S ARE
   TO BE THE SAME AS THAT OF THE PIPE.
   WELDS ON PIPING TO BE GROUND TO THE
   I.D. OF THE PIPE

Figure 3.10   Turbine meter installation.

7. A full-opening ball valve equal to or one size smaller than the turbine meter shall be installed on the upstream and downstream
   side of the meter. The inlet valve shall have a bypass for purging
   and pressuring the meter piping. 2-, 3-, 4-, 6-, and 8-in valves will
   have a ½-in bypass, and 10-in and larger valves will have a 1-in
   bypass.

8. A ½-in spacer shall be installed in the flange connection downstream of the meter to permit separation of the downstream meter
   spool from the turbine meter. Jackscrews shall be provided on the
   flanges at the spacer location.

9. For turbine meters 6 in and larger, a hoist shall be installed for spin testing and maintenance.

10. Flow-limiting orifices can be used in a turbine meter installation to prevent the flow from exceeding the capacity of the meter. Prolonged overranging of a turbine meter may result in damage of the rotor shaft bearings or destruction of the rotor itself.

The exact location of the orifice is not critical; however, the location must be in the downstream piping and should be from 5 to 10 pipe diameters from the meter outlet. With orifices to limit the flow through a turbine meter to 115 percent of the capacity of the meter, a pressure drop of 50 percent of the line pressure can be expected across the orifice plate. The flow through the orifice will reach critical flow when

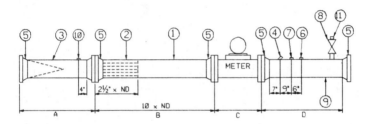

ND: NOMINAL DIAMETER (ie. 8")
ITEM SPECIFICATIONS

① GRADE B SEAMLESS PIPE   •I.D., •WAIT. •/FT.,W.P.(0.5 D.F.)   psig.
② TUBULAR STRAIGHTENING VANE (FLANGE TYPE)
③ BASKET STRAINER: 150% STRAINER OPENING WITH   ×   S.S.WIRE MESH.
④ 45° LATERAL CONNECTION, DRILLED FOR 1" SEPARABLE SOCKET (FOR THERMOMETER BULB OR R.T.D.)
⑤ WELDING NECK FLANGE, RAISED FACE, ANSI
⑥ ½" THREADOLET FOR BLOW-DOWN (FOR 2" TUBES USE¼" VALVE. FOR 3" TUBES USE½" VALVE).
⑦ ½" THREADOLET,½" NPT ⁷⁄₁₆" BORE, STAINLESS STEEL TEST WELL C/W CHAIN AND PLUG.
⑧ 2" F.O., S.E., BALL VALVE, ANSI
⑨ WELDING TEE SIZE   STD. OR X-HEAVY. USE REDUCING OUTLET FOR 3" TUBE.
⑩ ½" THREADOLET WITH SAMPLE PROBE. (1/2 " M ×¼" F PROBE).
⑪ 2" BULL PLUG,ASTM 105,ANSI B16.11

SIZE	ANSI	A	B	C	D
2"	300 600	1'-8"	1'-8"	SEE MANUFACTURER'S DRAWINGS	3'-0"
3"	300 600	1'-8"	2'-6"		3'-2"

NOTES:
1. INSTALL½" TREADOLET JUST U.S. OF U.S. ISOLATING VALVE.
2. FOR ANGLE BODY METER USE SAME U.S. AND D.S. SPOOLS AS SHOWN ABOVE
3. PIPE INTERIOR IS TO BE OF COMMERCIAL ROUGHNESS,AND THE FLANGE I.D.'S ARE TO BE THE SAME AS THAT OF THE PIPE. WELDS ON PIPING TO BE GROUND TO THE I.D. OF THE PIPE
4. LENGTH OF SEPARABLE SOCKET (ie. THERMOWELL) TO BE 3", EXCLUDING THREADS.

Figure 3.11   2- and 3-in turbine meter installation.

TABLE 3.5   Critical Flow through an Orifice or Choke Nipple

$Q = CP_1/G_rT_1$, scf/h

C = coefficient of orifice

P_1 = Upstream pressure lb/in^2 absolute

G_r = specific gravity of gas (air = 1.0)

T_1 = upstream temperature, °R

Orifice or nipple size, in		C value		
		Orifice		
		2-in pipe	4-in pipe	Choke nipple
$\frac{1}{16}$	0.063	1.524		
$\frac{3}{32}$	0.094	3.355		
$\frac{1}{8}$	0.125	6.301		6.25
$\frac{3}{16}$	0.188	14.47		14.44
$\frac{7}{32}$	0.218	19.97		20.0
$\frac{1}{4}$	0.250	25.86	24.92	26.51
$\frac{5}{16}$	0.313	39.77		43.64
$\frac{3}{8}$	0.375	56.58	56.01	61.21
$\frac{7}{16}$	0.438	81.09		85.13
$\frac{1}{2}$	0.500	101.8	100.2	112.72
$\frac{5}{8}$	0.625	154.0	156.1	179.74
$\frac{3}{4}$	0.750	224.9	223.7	260.99
$\frac{7}{8}$	0.875	309.3	304.2	355.0
1	1.000	406.7	396.3	475.0
$1\frac{1}{8}$	1.125	520.8	499.2	600.0
$1\frac{1}{4}$	1.250	657.5	616.4	750.0
$1\frac{3}{8}$	1.375	807.8	742.1	910.0
$1\frac{1}{2}$	1.500	1002.0	884.3	1100.0
$1\frac{3}{4}$	1.750		1208.0	1500.0
2	2.000		1596.0	2000.0
$2\frac{1}{4}$	2.250		2046.0	
$2\frac{1}{2}$	2.500		2566.0	
$2\frac{3}{4}$	2.750		3177.0	
3	3.000		3904.0	

the pressure drop across the orifice is 50 percent of the line pressure. Table 3.5 shows flow coefficients for various sizes of orifice or choke nipple.

Positive displacement meters. See Fig. 3.12 for a typical positive displacement meter installation. Refer to "Control—Pressure and Flow," Sec. 3.3.1E for overpressure protection requirements for positive meters. Positive meters (or rotary meters) are not to be used for production (wellhead) metering.

The capacity and dial rate of a positive displacement meter must be corrected in proportion to the flow pressure and the base pressure. For an Energy Economics HP3000 high-pressure positive meter operating at 7 inH$_2$O differential and measured flow Q = 3000 scf/h,

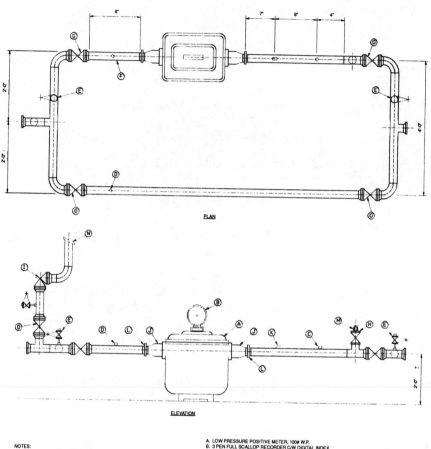

NOTES:

1. POSITIVE METER TO HAVE 1" N.P.T. TOP TAP FOR
   THERMOWELL WHEN AVAILABLE.
2. RELIEF VALVE IS NOT REQUIRED WHEN UPSTREAM MAOP IS ABOVE
   THE W.P. OF METER, EVEN IF U.S. REGULATION HAS MONITOR.
3. RELIEF VALVE IS NOT REQUIRED IF THE UPSTREAM MAOP IS
   BELOW THE W.P. OF METER.
4. A LOW PRESSURE POSITIVE METER IS CONSIDERED TO BE
   ANY METER WITH A W.P. LESS THAN OR EQUAL TO 500#.

A. LOW PRESSURE POSITIVE METER, 100# W.P.
B. 3 PEN FULL SCALLOP RECORDER C/W DIGITAL INDEX.
C. 1/2" THREADOLET WITH 1/2" N.P.T. TEST WELL. C/W CHAIN & PLUG.
D. 1/2" THREADOLET WITH SAMPLE PROBE (1/2" M X 1/2" F WITH PROBE).
E. 1/2" THREADOLET, 1/2" M X 1/4" F HAND VALVE, 1/4" BULL PLUG.
F. 2" NOMINAL PIPE.
G. 2" F.D., F.E., R.F., BALL VALVE.
H. 2" F.O., S.E., BALL VALVE.
I. RELIEF VALVE.
J. 4" X 2" N.P.T. SWAGE, XH.
K. 45 LATERAL CONNECTION WITH 1" THREADED OUTLET FOR
   1" N.P.T. THERMOWELL.
L. 2 C.S. PIPE UNION, CLASS 3000.
M. 2" BULL PLUG, ASTM 105, ANSI
N. PLASTIC CAP OR COVER
O. F.O. BALL VALVE.

Figure 3.12   Low-pressure positive meter station.

$$\text{Corrected } Q = (3000 \text{ scf/h})\left(\frac{\text{flow pressure, lb/in}^2 \text{ absolute}}{\text{base pressure, lb/in}^2 \text{ absolute}}\right)$$

Capacity must also be derated in order to control the rotational speed
of internal parts. The following derating factors apply:

0–250 lb/in^2 gauge   Derate by 20%

251–750 lb/in^2 gauge   Derate by 30%

751–1200 lb/in^2 gauge   Derate by 40%

1201–1500 lb/in^2 gauge   Derate by 50%

Example   Measuring pressure is 800 lb/in^2 gauge. Then

$$\text{Corrected } Q = (3000 \text{ scf/h}) \left(\frac{800 + 14.7}{14.73} \right) = 165{,}927 \text{ scf/h}$$

where 14.7 = assumed atmospheric pressure, lb/in^2. Then for 800 lb/in^2, derate by 40 percent. Therefore

$$165{,}927(0.60) = 99{,}556 \text{ scf/h}$$

Thus, 99,556 scf/h is the maximum capacity at 800 lb/in^2 gauge.

Now to find the proper size of the critical flow orifice for the open rate. First find maximum hourly dial rate capacity.

$$\text{Pressure factor} = \frac{\text{flow pressure, lb/in}^2 \text{ absolute}}{\text{base pressure, lb/in}^2 \text{ absolute}}$$

$$= \frac{800 + 14.7}{14.73} = 55.309$$

Then

$$\text{Max. hourly dial rate capacity} = \frac{\text{Max. capacity at flow pressure}}{\text{pressure factor}}$$

$$= \frac{99{,}556 \text{ scf/h}}{55.309} = 1800 \text{ cfh}$$

Annubar meters. An Annubar is a primary flow sensor designed to produce a differential pressure proportional to flow. Annubar meters may be installed in special cases for noncustody transfer application. Annubar meters shall be Flo Tap type. Refer to Fig. 3.13.

Vortex meters. Vortex meters may be installed in special cases in noncustody transfer applications. Small (1-in nominal diameter) vortex meters sometimes require standard wall thickness pipe on the meter inlet. Therefore, special approval must be obtained for installation of these meters.

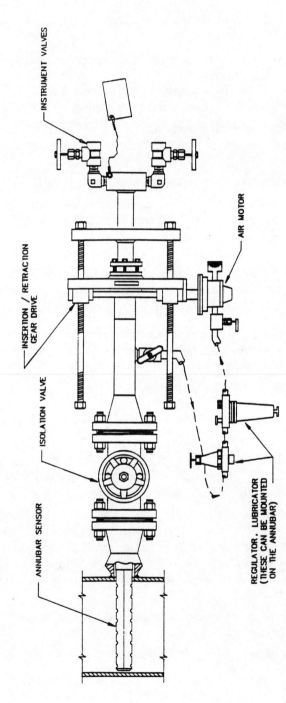

INSTRUMENT VALVES

AIR MOTOR

INSERTION / RETRACTION GEAR DRIVE

ISOLATION VALVE

ANNUBAR SENSOR

REGULATOR, LUBRICATOR
(THESE CAN BE MOUNTED ON THE ANNUBAR)

Figure 3.13 Annubar meter.

D. Offshore applications

1. Topics not specifically addressed here are covered under "General piping rules," Sec. 3.3.1A.

2. Offshore meter stations are usually designed as skid-mounted dual meter runs. Piping on meter skids shall be designed, constructed, and tested in accordance with ANSI B31.3, "Code for Chemical Plant and Petroleum Refinery Piping."

3. Minimum length of straight pipe shall be as specified by AGA Gas Measurement Committee Report No. 3, revised 1985, for a β ratio of 0.75.

4. Diameter of headers shall be as specified in Table 3.4.

5. Carefully selected smooth steel pipe, ASTM A-106, Grade B, shall be used for meter tubes. All other piping shall be ASTM A-106, Grade B seamless. AP1 5L Grade B seamless may be substituted if A-106 is not available.

6. Setscrew-lock-type straightening vanes shall be installed where dehydration facilities have been provided on the platforms.

7. Meter runs shall be equipped with taps, as specified on Fig. 3.7.

8. 2-in-orifice fittings, senior-type, shall be installed on the meter's side. Each meter run shall be equipped with valves upstream and downstream of the orifice fitting.

9. Meter run valves will be full-opening ball valves and the same size as the meter run. Valves 6 in and larger are to be gear-operated. Ball valves shall have $\frac{1}{2}$-in-body bleeder valves.

10. Flanges ANSI 600 and above shall be RTJ (ring type joint). Flanges ANSI 400 and lower shall be RF (raised face) with a serrated spiral finish. The senior-type orifice fitting shall be tongue-and-groove type with a flat gasket.

11. Meter houses shall be walk-in type. Panels may be fiberglass. Steel framing members shall be galvanized steel. Stainless steel bolts shall be used. Adequate ventilation must be provided to satisfy U.S. Department of the Interior, O.C.S. (Outer Continental Shelf) orders, in lieu of automatic gas detection, on offshore platforms.

12. A turbine meter may be used under special circumstances of space and gas quality.

13. Liquid and gas will normally be separated and measured separately.

E.   Control—pressure and flow.   Refer also to "Electronic measurement and control," Sec. 3.3.2D for alternative control hardware, telemetering equipment, and run switching equipment. Control requirements are dictated by the needs of the customer or supplier and by the gas control method, and therefore specific applications are not given.

Pipeline delivery applications (sales/transport delivery)

A.   General
1. In general, regulators (primary and monitor) will be designed to fail open with loss of output signal from controller.
2. All valves and piping are to be designed to withstand mainline delivery pressure.
3. Overpressuring shall be protected against by a monitoring regulator installed in series with the primary regulator. The monitor is to be installed downstream of the regulator. Each regulator must have an independent control and supply system.
4. Standby regulator run capacity shall be installed at critical locations. In a dual-run design, each run shall be capable of carrying the entire station load. If more than two regulator runs are required, the entire load shall be carried by all runs less one.
5. Regulator runs shall be installed with upstream and downstream shutoff valves in each regulator run. The design of the regulator setting shall be such that any single incident, such as an explosion or damage by a vehicle, will not affect the operation of both the overpressure protection device and the primary regulator.
6. Meter stations with a positive meter, not rated for mainline pressure, will have, in addition to the monitoring regulator, a relief valve solely for the protection of the meter case. Piping on the outlet of the relief valve is to be one size larger than the inlet.

B.   Design features
1. Conventional regulators such as the Grove and Fisher Whisper Trim, in sizes 1 through 6 in, and the Jet Stream in sizes 1, 2, and 4 in only, should be used as primary regulators on applications that require capacities in their range on single-, dual-, and triple-run stations, requiring pressure drops in the 50- to 800-lb/in^2 gauge range. Monitor regulators should be of the conventional hard seat type, i.e., Fisher with E body. Flexible-sleeve-type regulators (such as Grove Flex Flo) may be used as primary as well as monitor regulators. For eco-

nomic reasons, an isolating valve on a regulator run is also used as a monitor at times.

2. Large-capacity, low-differential regulators shall be of the valve regulator type. Normally the minimum size valve regulator shall be 4 in and shall not exceed 16 in. Piping configuration shall include a bypass, where necessary.

3. All regulator piping is to be designed with 0.5 design factor and sized to restrict flowing velocities to 100 ft/s abovegrade and 200 ft/s belowgrade to eliminate as much noise, vibration, and erosion as possible. Depending on location and allowable noise, a low-noise regulator (such as Neles Q-ball) may have to be used.

4. Changes in regulator piping shall be made adjacent to the control valve by using concentric reducers to reduce velocity and increase capacity. A 2-pipe-diameter size increase is considered nominal.

5. All regulators 4 in and larger shall be adequately anchored with piers and hold-downs, both upstream and downstream of the regulators, to reduce vibration.

6. To avoid vibration damage, controllers shall not be mounted on the valve body or associate piping. Instrument columns, when installed, shall be anchored to the building foundation as far away from the pipeline as possible, but not less than 4 ft, to eliminate excessive vibration on the instrumentation. Sensing line taps shall be installed out of the turbulent area, a minimum of 15 ft downstream of the regulator or on the bypass run, if available. Figures 3.14, 3.15, and 3.16 show schematics of control piping and control hookups for various configurations.

7. All instrument supply regulators shall be provided with sufficient heating to prevent hydrocarbon condensation in the instrument gas (refer to Fig. 3.17).

8. At measuring stations where automatic station bypasses are provided, the downstream valve shall be of the sliding disk type. The objective is to obtain a positive shutoff.

Pipeline receipt applications (purchase/transport receive)

A. *General:* Receipt meters requiring flow control with pressure override protection shall have the primary regulator downstream of the orifice meter. The monitor regulator may be installed upstream of the orifice meter. Each regulator shall have separate sensing and supply pressure points.

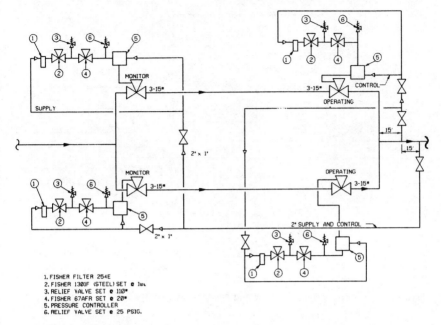

1. FISHER FILTER 254E
2. FISHER 1301F (STEEL) SET @ 10ʷ
3. RELIEF VALVE SET @ 110ʷ
4. FISHER 67AFR SET @ 20ʷ
5. PRESSURE CONTROLLER
6. RELIEF VALVE SET @ 25 PSIG.

Figure 3.14   Regulator—control piping—pressure control.

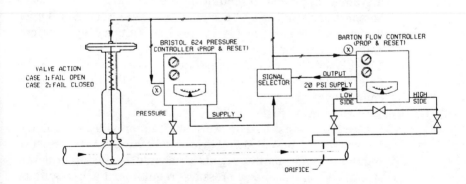

NOTES: 1. ⓧ CONNECTION TO CONTROLLER RESET BELLOWS INPUT
         2. CONTROL SYSTEM ACTIONS:

CASE NO.	VALVE ACTION	FLOW OR PRESSURE CONTROLLER	SIGNAL SELECTED
1	FAIL OPEN	REVERSE	HIGHEST
2	FAIL CLOSED	DIRECT	LOWEST

Figure 3.15   Control valve with flow and pressure controller hookup.

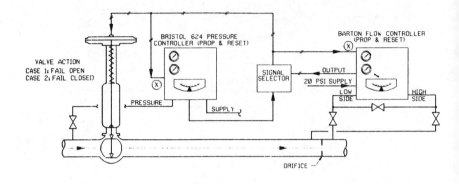

NOTES:

1. ⊗ CONNECTION TO CONTRLLER RESET BELLOWS INPUT
2. CONTROL SYSTEM ACTIONS

CASE NO.	VALVE ACTION	FLOW CONTROLLER	BACK PRESSURE CONTROLLER	SIGNAL SELECTED
1	FAIL OPEN	DIRECT	REVERSE	HIGHEST
2	FAIL CLOSED	REVERSE	DIRECT	LOWEST

Figure 3.16 Control valve with flow and backpressure controller.

B. Design features

1. Flow control with pressure override can be achieved by using a systems approach to regulating control. Installation should be carefully studied and designed for the application. Other design features applicable to pipeline delivery applications stated earlier apply.

2. The Fisher Wizard controller offers a cost savings over the Bristol 624, but should be used only where the line conditions change very slowly.

C. Compressor station fuel

1. Protection against overpressure shall be provided by a monitor regulator, installed in series with the primary pressure regulator. The monitor regulator should be installed downstream of the primary pressure regulator. Each regulator shall have separate sensing and supply pressure points located downstream of the primary pressure regulator.

2. The primary regulator shall close with an increase in signal and open with a loss of signal.

3. The monitor regulator shall open with an increase in signal and close with a loss of signal.

4. A standby regulator run or bypass shall be installed as appropriate.

Overpressure protection

A. Requirements
1. For receipt applications (purchase, exchange, etc.) over-pressure protection instrumentation shall be installed to protect facilities from pressures exceeding the lower of two limits:
   a. 110 percent MAOP
   b. Pressure which produces a hoop stress of 75 percent SMYS (specified minimum yield strength).
2. For delivery applications (sales, transportation, etc.), over-pressure protection instrumentation shall be installed to pro-

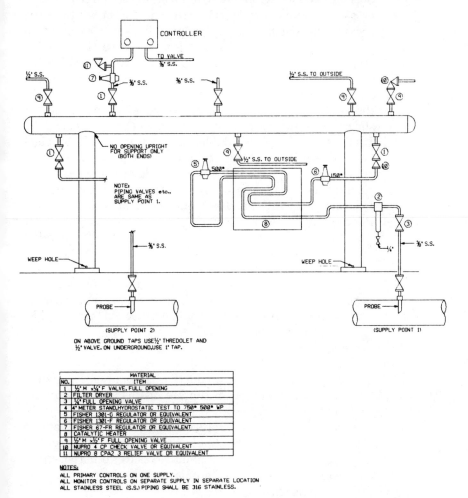

ON ABOVE GROUND TAPS USE ½" THREDOLET AND ½" VALVE. ON UNDERGROUND, USE 1" TAP.

MATERIAL	
NO.	ITEM
1	½" M x ¼" F VALVE, FULL OPENING
2	FILTER DRYER
3	¼" FULL OPENING VALVE
4	4" METER STAND, HYDROSTATIC TEST TO 750° 500° WP
5	FISHER 1301-G REGULATOR OR EQUIVALENT
6	FISHER 1301-F REGULATOR OR EQUIVALENT
7	FISHER 67-FR REGULATOR OR EQUIVALENT
8	CATALYTIC HEATER
9	½" M x½" F FULL OPENING VALVE
10	NUPRO 4 CP CHECK VALVE OR EQUIVALENT
11	NUPRO 8 CPA2 3 RELIEF VALVE OR EQUIVALENT

NOTES:
ALL PRIMARY CONTROLS ON ONE SUPPLY.
ALL MONITOR CONTROLS ON SEPARATE SUPPLY IN SEPARATE LOCATION
ALL STAINLESS STEEL (S.S.) PIPING SHALL BE 316 STAINLESS.

Figure 3.17   Instrument gas supply piping.

tect facilities from pressures exceeding maximum allowable operating pressure or maximum allowable delivery pressure as a result of equipment malfunction or other abnormal condition.

B. Equipment
   1. Relief valves
      a. For meters measuring dry gas, a relief valve shall be installed to protect the meter facilities, where required.
      b. Relief protection shall be provided wherever the inlet pressure exceeds the outlet pressure rating of the regulator.
      c. Specification for relief valves shall state the relief, or "pop," pressure and the reseat pressure. The pop pressure is the pressure at which the relief valve will open (on increasing pressure) and begin discharging to decrease the pipeline pressure. The reseat pressure is the pressure at which the relief valve will close (on decreasing pressure) once enough gas has been discharged to lower the pipeline pressure to the reseat pressure setting of the relief valve. Table 3.6 shows capacities of nozzle-type safety relief valves.
      d. Dual ports, where available, shall be specified for side-vented relief valves larger than 1½ in, to prevent excessive side thrust when the valve relieves. Side thrust must particularly be considered when pop pressures in the 1000 lb/in^2 range are realized. Braces are required.
      e. Pilot-operated relief valves shall be equipped with a pilot line filter and a no-flow modulating pilot. Where relief valves are used in wet gas service, a dehydrator shall also be installed in the pilot line.
      f. Field test connections must be provided on relief valves to permit periodic in-service relief valve checks (Fig. 3.18). A shutoff valve equipped so that it can be locked open shall be installed under each relief valve.
   2. Rupture disks
      a. General
         i. In remote areas on production meters and for meters measuring wet gas, a rupture disk is installed to protect the meter facilities.
         ii. Manufacturers of rupture disks recommend that the operating pressure should be less than 90 percent of the rupture disk's set point pressure; otherwise, the rupture disk will fail prematurely.
      b. Design considerations
         i. Regardless of the manufacturer, it is always recommended that only disks with "zero" manufacturing

TABLE 3.6   Gas Capacities of Nozzle-Type Safety Relief Valves
(Thousands of Cubic Feet Per Hour), Specific Gravity 0.60, Temperature 60°F, 5% Accumulation

Set pressure, lb/in² gauge	Orifice letter and area, in²*											
	D	E	F	G	H	J	K	L	N	P	Q	R
	0.110 (0.16)	0.196 (0.29)	0.307 (0.45)	0.5027 (0.74)	0.7854 (1.15)	1.2868 (1.89)	1.8385 (2.70)	2.8526 (4.19)	4.34 (6.37)	6.3794 (9.37)	11.045 (16.21)	16.00 (23.50)
10	3.9	6.9	10.8	17.7	28	46	65	101	153	225	389	563
20	5.5	9.8	15.3	24.6	39	64	92	142	217	318	551	798
30	7.1	12.7	19.8	32.5	51	83	119	184	280	412	713	1,034
40	8.7	15.5	24.3	39.8	62	102	146	226	344	506	875	1,269
50	10.3	18.4	28.8	47.2	74	121	173	268	408	600	1,037	1,503
60	12.0	21.3	33.3	54.6	85	140	200	310	471	693	1,199	1,738
70	13.6	24.2	37.8	62.0	97	159	227	352	535	787	1,362	1,973
80	15.2	27.0	42.4	69.4	108	178	254	394	599	880	1,524	2,208
90	16.8	29.9	46.9	76.7	120	196	281	435	662	974	1,686	2,442
100	18.4	32.8	51.4	84.1	131	215	308	477	726	1,067	1,848	2,677
110	20.0	35.7	55.9	91.5	143	234	335	519	790	1,161	2,010	2,913
120	21.6	38.5	60.4	98.9	154	253	366	561	853	1,255	2,172	3,148
130	23.2	41.4	64.9	106.2	166	272	389	603	917	1,348	2,334	3,382
140	24.7	44.3	69.4	113.6	178	291	416	645	981	1,442	2,496	3,617
150	26.5	47.2	73.9	121.0	189	310	442	687	1,045	1,535	2,658	3,852
160	28.1	50.1	78.4	128.4	201	329	470	728	1,108	1,629	2,820	4,087
170	29.7	52.9	82.9	135.7	212	348	496	770	1,172	1,723	2,982	4,321
180	31.3	55.8	87.4	143.1	224	366	523	812	1,236	1,816	3,144	4,556
190	32.9	58.7	91.9	150.5	235	385	550	854	1,299	1,910	3,307	4,791
200	34.5	61.6	96.4	157.9	247	404	577	896	1,363	2,004	3,469	5,027
210	36.2	64.4	100.9	165.3	258	423	604	938	1,427	2,097	3,631	5,261
220	37.8	67.3	105.4	172.6	270	412	632	980	1,490	2,191	3,793	5,496
230	39.7	70.2	109.9	180.0	281	461	658	1,021	1,554	2,284	3,955	5,730
240	41.0	73.1	114.4	187.4	293	480	685	1,063	1,618	2,378	4,117	5,966
250	42.6	75.9	118.9	194.8	304	499	712	1,105	1,681	2,472	4,279	6,200
260	44.2	78.8	123.4	202.1	316	517	739	1,147	1,745	2,565	4,441	6,435
270	45.8	81.7	128.0	209.5	327	536	766	1,189	1,809	2,659	4,603	6,671
280	47.5	84.6	132.5	216.9	339	555	793	1,231	1,872	2,753	4,765	6,906
290	49.1	87.4	137.0	224.3	350	574	820	1,273	1,936	2,846	4,928	7,140
300	50.7	90.3	141.5	231.6	361	593	847	1,315	2,000	2,940	5,090	7,375
310	52.3	93.2	146.0	239.0	373	612	874	1,356	2,064	3,033	5,252	7,610
320	53.9	96.1	150.5	246.4	385	631	901	1,398	2,127	3,127	5,414	7,845
330	55.5	99.0	155.0	253.8	397	650	928	1,440	2,191	3,221	5,576	8,088
340	57.2	101.8	159.5	261.2	408	669	995	1,482	2,255	3,314	5,738	8,314
350	58.8	104.7	164.0	268.5	420	687	982	1,524	2,318	3,407	5,900	8,549
360	60.4	107.6	168.5	275.9	431	706	1,009	1,566	2,382	3,502	6,062	8,784
370	62.0	110.5	173.0	283.3	442	725	1,036	1,608	2,446	3,595	6,224	9,019
380	63.6	113.3	177.5	290.7	454	744	1,063	1,649	2,509	3,689	6,386	9,253
390	65.2	116.2	182.0	298.0	466	763	1,090	1,691	2,573	3,782	6,548	9,488
400	66.8	119.1	186.5	305.4	477	782	1,117	1,733	2,637	3,876	6,710	9,723
410	68.5	122.0	191.0	312.8	489	801	1,144	1,775	2,701	3,970	6,873	9,959
420	70.1	124.8	195.5	320.2	500	820	1,171	1,817	2,764	4,063	7,035	10,194
430	71.7	127.7	200.0	327.5	512	839	1,198	1,859	2,828	4,157	7,197	10,428
440	73.3	130.6	204.5	334.9	523	857	1,225	1,901	2,892	4,250	7,359	10,663
450	74.9	133.5	209.1	342.3	535	876	1,252	1,942	2,955	4,345	7,521	10,898

TABLE 3.6   Gas Capacities of Nozzle-Type Safety Relief Valves (*Continued*)

Set pressure, lb/in² gauge	Orifice letter and area, in²*											
	D	E	F	G	H	J	K	L	N	P	Q	R
	0.110	0.196	0.307	0.5027	0.7854	1.2868	1.8385	2.8526	4.34	6.3794	11.045	16.00
	(0.16)	(0.29)	(0.45)	(0.74)	(1.15)	(1.89)	(2.70)	(4.19)	(6.37)	(9.37)	(16.21)	(23.50)
460	76.5	136.3	213.6	349.7	546	895	1,279	1,984	3,019	4,438	7,683	11,133
470	78.1	139.2	218.1	357.1	558	914	1,306	2,026	3,083	4,531	7,845	11,367
480	79.8	142.1	222.6	364.4	569	933	1,333	2,068	3,146	4,625	8,007	11,602
490	81.4	145.0	227.1	371.8	581	952	1,360	2,110	3,210	4,719	8,169	11,837
500	83.0	147.8	231.6	379.2	593	971	1,387	2,152	3,274	4,812	8,331	12,072
510	84.6	150.7	236.1	386.6	604	980	1,414	2,194	3,337	4,906	8,492	12,304
520	86.2	153.6	240.6	393.9	616	1,008	1,441	2,235	3,401	5,000	8,655	12,540
530	87.8	156.5	245.1	401.3	627	1,027	1,468	2,277	3,465	5,094	8,818	12,776
540	89.4	159.4	249.6	408.7	639	1,046	1,495	2,319	3,528	5,187	8,979	13,009
550	91.0	162.2	254.1	416.1	650	1,065	1,522	2,361	3,592	5,280	9,140	13,242
560	92.7	165.1	258.6	423.5	662	1,084	1,549	2,403	3,656	5,374	9,302	13,478
570	94.3	168.0	263.1	430.8	673	1,103	1,576	2,445	3,719	5,468	9,465	13,714
580	95.9	170.9	267.6	438.2	685	1,122	1,603	2,487	3,783	5,561	9,626	13,947
590	97.5	173.7	272.1	445.6	696	1,141	1,630	2,529	3,847	5,665	9,789	14,183
600	99.1	176.6	276.6	453.0	708	1,160	1,657	2,570	3,911	5,749	9,952	14,418
610	100.7	179.5	281.1	460.3	719	1,178	1,684	2,612	3,974	5,842	10,113	14,651
620	102.3	182.4	285.6	467.7	731	1,197	1,711	2,654	4,038	5,936	10,275	14,888
630	104.0	185.2	290.1	475.1	742	1,216	1,738	2,696	4,102	6,030	10,438	15,123
640	105.6	188.1	294.7	482.5	754	1,235	1,765	2,738	4,165	6,123	10,599	15,356
650	107.2	191.0	299.2	489.8	765	1,254	1,792	2,780	4,229	6,217	10,762	15,592
660	108.8	193.9	303.7	497.2	777	1,273	1,819	2,822	4,293	6,310	10,923	15,825
670	110.4	196.7	308.2	504.6	788	1,292	1,845	2,863	4,356	6,404	11,085	16,061
680	112.0	199.6	312.7	512.0	800	1,311	1,872	2,905	4,420	6,498	11,248	16,297
690	113.6	202.5	317.2	519.3	811	1,330	1,899	2,947	4,484	6,591	11,409	16,530
700	115.3	205.4	321.7	526.7	823	1,348	1,926	2,989	4,548	6,685	11,572	16,766
710	116.9	208.3	326.2	534.1	835	1,367	1,953	3,031	4,611	6,778	11,733	16,999
720	118.5	211.1	330.7	541.5	846	1,386	1,980	3,073	4,675	6,872	11,895	17,235
730	120.1	214.0	335.2	548.8	858	1,405	2,007	3,115	4,738	6,965	12,056	17,468
740	121.7	216.9	339.7	556.3	869	1,424	2,034	3,156	4,802	7,059	12,219	17,704
750	123.3	219.8	344.2	563.6	881	1,443	2,061	3,198	4,866	7,153	12,382	17,940
760	124.9	222.6	348.7	571.0	892	1,462	2,088	3,240	4,930	7,247	12,545	18,175
770	126.6	225.5	353.2	578.4	904	1,481	2,115	3,282	4,993	7,340	12,706	18,409
780	128.1	228.4	357.7	585.8	915	1,499	2,142	3,324	5,057	7,434	12,868	18,644
790	129.8	231.3	362.2	593.1	927	1,518	2,169	3,366	5,121	7,528	13,031	18,880
800	131.4	234.1	366.7	600.5	938	1,537	2,206	3,408	5,184	7,621	13,192	19,113
850	139.5	248.5	389.3	637.4	996	1,632	2,331	3,617	5,503	8,089	14,002	20,287
900	147.5	262.9	411.8	674.3	1,054	1,726	2,466	3,826	5,821	8,558	14,814	21,463
950	155.6	277.3	434.5	711.2	1,111	1,820	2,601	4,036	6,143	9,037	15,627	22,665
1000	163.7	291.7	456.8	748.0	1,169	1,915	2,736	4,245	6,461	9,505	16,437	23,839
1050	171.8	306.0	479.3	784.9	1,226	2,009	2,871	4,454				
1100	179.8	320.4	501.9	821.8	1,284	2,104	3,006	4,663				
1150	187.9	334.8	524.4	858.7	1,342	2,198	3,140	4,873				
1200	195.9	349.2	546.9	895.6	1,399	2,293	3,275	5,082				

*Values in parentheses give approximate increase in gas capacity for each 1 lb/in² increase in set pressure.

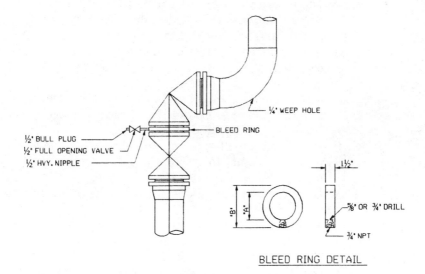

BLEED RING DETAIL

NOTE: BLEED RING MATERIAL TO BE CARBON STEEL UNLESS OTHERWISE DESIGNATED.
CONNECT TEST VALVE TO POINT UPSTREAM OF REGULATOR WITH 1" PIPE.

DIMENSIONS IN INCHES

SIZE	'A' FOR R.F.	'B' FOR R.F.			
		150 LB.	300 LB.	600 LB.	900 LB.
1	$1\frac{1}{8}$	$2\frac{1}{2}$	$2\frac{3}{4}$	$2\frac{3}{4}$	3
$1\frac{1}{2}$	$1\frac{7}{8}$	$3\frac{1}{4}$	$3\frac{5}{8}$	$3\frac{5}{8}$	$3\frac{3}{4}$
2	$2\frac{1}{8}$	4	$4\frac{1}{4}$	$4\frac{1}{4}$	$5\frac{1}{2}$
$2\frac{1}{2}$	$2\frac{1}{2}$	$4\frac{3}{4}$	5	5	$6\frac{3}{8}$
3	$3\frac{1}{8}$	$5\frac{1}{4}$	$5\frac{3}{4}$	$5\frac{3}{4}$	$6\frac{1}{2}$
4	$4\frac{1}{8}$	$6\frac{3}{4}$	7	$7\frac{1}{2}$	8
6	$6\frac{1}{8}$	$8\frac{5}{8}$	$9\frac{3}{4}$	$10\frac{3}{8}$	$11\frac{1}{4}$
8	8	$10\frac{7}{8}$	12	$12\frac{1}{2}$	14
10	$10\frac{1}{8}$	$13\frac{1}{4}$	$14\frac{1}{8}$	$15\frac{5}{8}$	17
12	12	16	$16\frac{1}{2}$	$17\frac{3}{4}$	$19\frac{3}{8}$

Figure 3.18   Bleed ring for relief valve testing.

range be used. This significantly reduces the chance of fatigue failure when the actual operating pressure approaches a pipeline's MAOP.

ii. The ASME (American Society of Mechanical Engineers) Boiler and Pressure Vessel Code, Sec. VIII, Division I, which the DOT refers to, states: Every rupture disk shall be guaranteed by its manufacturer to burst within ±5 percent of its stamped bursting pressure at coincident temperature.

iii. 316 stainless steel disks will satisfy most needs.

iv. The disk's holder is available in either stainless steel or carbon steel. The cost of carbon steel is considerably less and is sufficient for most applications.

   v. The holder has a flat seat to match the flat-seat disk. Some manufacturer's holders do not accept other companies' disks.

   vi. A fail-safe disk is one that, if damaged or installed incorrectly, will burst at or below the stamped burst pressure.

   vii. Insert holders are placed between two companion flanges. Flanges should be threaded type if the smaller sizes are used.

   viii. A plastic blow-off–type cap placed on top of the outlet pipe is necessary to protect the disk from the environment.

   ix. Each location should have at least one extra disk as an emergency spare.

   x. An isolating valve should be installed ahead of the rupture disk.

   xi. The rupture disk should be placed just upstream of the meter run's upstream valve at a receiving point.

 c. Installation and handling

   i. Rupture disks are made of very thin material and are extremely fragile. Therefore much care must be taken not to touch the bubble of the disk.

   ii. A small tap and hand valve should be installed between the isolating valve and the rupture disk. This is to blow down this section of pipe before inspection.

   iii. One must ensure that the disk is not installed upside down. Stamped side of handle faces up. Also, holders have arrows pointing in the direction of flow.

   iv. Visual inspection should be performed annually. The rupture disk should be isolated from pressure and the plastic blow-off cap removed. One should then look for signs of potential problems. The blow-off cap should be replaced with a new one.

   v. When a rupture disk is placed back in service the isolating valves should be opened *very* slowly.

 d. Calculation of stamped burst pressure

   i. For overpressure protection equipment, DOT allows 10 percent over the MAOP of the line we are protecting.

   ii. For purposes of clarification, the following example is given.

Example A meter station is located *next* to a transmission line. MAOP is 780 lb/in^2. Contract flow rate from the producer is 6,000,000 scf/day. Minimum operating pressure of the transmission line is 500 lb/in^2.

$$780 + 10\% = 780 + 78 = 858 \text{ lb/in}^2 \quad \text{(DOT max.)}$$

Burst tolerance is $\pm 5\%$, so

$$P \times 1.05 = 858$$

$$P = 817$$

Therefore, 817 lb/in^2 is the stamped burst pressure desired.
Now to find safe operating range for the disk:

$$P = 817 - (0.05 \times 817)$$
$$= 776$$

So the disk will not burst at less than 776 lb/in^2. However, the disk's operating range is 0 to 90 percent of the stamped burst pressure:

$$817 (0.90) = 735$$

This means that the transmission line should not be operated at pressures exceeding 735 lb/in^2 in order to ensure that rupture disks do not fail prematurely due to fatigue. If a rupture disk application may lower the capacity of the line by lowering MAOP, then a rupture disk may not be the right choice for the application.

    iii. For cases where the meter station and transmission line are separated by gathering system lines, the calculations become more detailed and must be reviewed carefully for each location.

    iv. When pipelines are upgraded or derated, rupture disks must be reevaluated and replaced if necessary.

F. Gas sampling. Determination of gas composition by taking a representative sample and analyzing it is essential for accurate volume as well as energy calculation. To secure a true representative sample of natural gas at various locations, sampling facilities shall be provided.

1. A sample probe for meter run installations consists of a ¼-in male × female valve which is screwed into a ¾-in NPT (National Pipe Thread) × ¼-in NPT probe. The probe length shall be equal to approximately one-third the meter run diameter.

2. A sample probe for other locations may consist of a gate or ball valve on a short nipple through which heavy-wall ¼-in stainless steel tubing or pipe can be inserted into the pipeline. This probe is to be of sufficient length to protrude well away from the internal meter tube skin to avoid contamination from migrating heavy hydrocarbons, water, and solids. The sample line itself is normally ⅛-in stainless steel and shall be as short as practical. Pressure reduction, if required, should be located close to the probe to elim-

inate unnecessary lag time. Heat trace or catalytic heat may be required on the regulator sample line or instrument case itself, where hydrates or low atmospheric temperatures cause problems.

3. For most gases, a high-pressure sampler is preferred. The gas is sampled at line pressure and moved into a displacement-type cylinder that is also at line pressure. The gas pressure is not cut in the field and fallout is held to a minimum. This sampler can operate either from time impulse or flow rate signals.

4. Generally, any meter station which is expected to flow 50 MMscf/ day and above will require the installation of a gas chromatograph. If so, all aboveground sample tubing must be heat-traced to avoid possible condensation of water molecules or distillate formation which may occur if the ambient temperature is below the flowing gas temperature. If sampling lines of excessive length cannot be avoided, a constant purge of the line is mandatory. This will eliminate unnecessary lag time and prevent the sampling of a stagnant gas pocket.

3.3.2 Instrumentation (secondary measurement devices)

Certain secondary devices are required to convert the physical variables (differential pressure, static pressure, temperature, etc.) through the primary devices (such as an orifice tube) into some recorded code that can be then decoded into accounting information for custody transfer purposes.

The following are some of the essential elements of measurement instrumentation.

A. Flow recorder

1. The recorder shall be a two-pen (or three-pen if temperature is also desired on the same chart) type, measuring differential pressure in inches of water and static pressure in pounds per square inch absolute.

2. The recorder will use a dry bellows with temperature compensation and a linear chart. 24-hour chart rotation is recommended for meter stations with flow at 10 MMscf/day and higher. Weekly charts may be used for flow rates below 10 MMscf/day. However, test frequencies may be dictated by contract. Stations with flow less than 10 MMscf/day are tested quarterly and stations flowing more than 10 MMscf/day are tested monthly.

3. The Bourdon tube and range springs shall be selected so that the recorder will measure between 10 and 90 percent of scale.

4. Static pressure shall be obtained from the downstream pressure tap of the orifice fitting for flange tap connections. With pipe tap connections, pressure is obtained from the upstream pressure tap.

5. A standard five-valve manifold shall be installed along with valves at the pressure taps to facilitate testing and operating the recorder.

6. Gauge lines should be sloped at least ⅛ in/ft with the low point at the pressure taps. If an adequate slope is not possible, an inline drip/filter shall be installed.

7. The flow recorder for positive displacement meters should be a three-pen type that records pressure, volume, and temperature.

8. All tubing, fittings, and valves shall be stainless steel. See checklist in Fig. 3.19 for ordering descriptions for three-pen orifice, positive displacement, and turbine meter recorders.

B.   Temperature recorder (separate instrument or integral to flow recorder)

1. The temperature recorder shall be a capillary type with the same size and rotation as the flow recorder.

2. Under normal operating conditions, the temperature range should be 0 to 150°F.

3. The thermometer bulb shall be installed in a thermowell at the first threadolet downstream of the orifice plate (or downstream of the downstream pipe tap).

4. A test thermowell shall be installed approximately 6 in downstream of the recorder bulb.

C.   Gas sampling

1. A continuous sampler shall be used whenever feasible. If possible, this sampler should sample proportionally to flow.

2. The continuous sampler shall use a pitot-type sampling probe with a "speed loop." The distance between the sampler and sample probe should be minimized.

3. To obtain a representative gas sample, the gas sampling probe should be installed in the least turbulent section of meter station piping, preferably at the end of a straight run of pipe. If placed in a horizontal run of pipe, the probe should be inserted from the top.

Figure 3.19   Checklist for ordering descriptions for three-pen recorders.

I. **Three-Pen Recording Orifice Meters**
  A. Manufacturer and model number (All are bellows-type, 12-in chart size)
    ☐ 2-in pipe stand mount   ☐ wall mount   ☐ flush mount
    1. ☐ American Dri-Flo II, model 25FS2
    2. ☐ Barton model 202E,
    3. ☐ Barton model 202N (for sour gas service)
    4. ☐ Foxboro model 39AFPTR
    5. ☐ Other, _____
  B. Case
    ☐ Standard (glass in door)   ☐ Plexiglas window   ☐ solid door
  C. Pressure element material
    ☐ 316 stainless steel, helical coil
  D. Pressure element range, lb/in^2
    ☐ 0–250   ☐ 0–500   ☐ 0–1000   ☐ Other, 0– _____
  E. Differential range, inches WC (Water Column)
    ☐ 0–100   ☐ 0–200   ☐ Other, 0– _____
  F. Temperature range, °F
    ☐ 0–150 Other,
    ☐ 0– _____
  G. Chart rotation
    ☐ 24-hour   ☐ 7-day   ☐ 31-day
    ☐ Multiple rotation selection (must use battery-operated programmable chart drive)
    ☐ Other, _____
  H. Chart drive
    ☐ Conventional spring-wound
    ☐ Battery-operated, programmable (Wilson or Reynolds)
    ☐ Other, _____
  I. Inking system
    ☐ Bucket-type pens
    ☐ Disposable fiber-tip pens
    ☐ Other, _____
  J. Temperature system
    ☐ Adjustable Mercury bulb with 1½-in NPT sliding union adaptable to thermowell (separate socket), 1-in NPT separable socket to fit 1-in threadolet, insertion length of thermowell (excluding threads) _____ inches, length of stainless steel case-compensated capillary _____ ft (if greater than 25 ft, use fully compensated system.)
  K. Automatic chart changer
    ☐ Yes. If "yes," check off either battery-operated or conventional spring-wound in part *H* and check off 24-hour chart rotation in part *G*. Also, the recorder must have slot cut in bottom and be plated to accept Mullings automatic chart changer.
    ☐ No

II. **Three-Pen Recording Full Scallop Meters for Positive Displacement and Turbine Meters**
  A. Manufacturer and model number (all are 12-in chart size)
    1. ☐ Mercury model 1238
    2. ☐ Other, _____
  B. Case
    ☐ Standard (glass in door)
    ☐ Plexiglas window
    ☐ solid door

C. Pressure element material
- ☐ 316 stainless steel, helical coil

D. Pressure element range, lb/in^2
- ☐ 0–250 ☐ 0–500 ☐ 0–1000 ☐ Other, 0– _____

E. Temperature range, °F
- ☐ 0–150
- ☐ Other, 0– _____

F. Chart drive
- ☐ Conventional spring-wound
- ☐ Battery-operated
- ☐ Single-speed (standard with Mercury) ☐ dual-speed
- ☐ Battery-operated, programmable (Wilson or Reynolds)

G. Chart rotation
- ☐ 24-hour ☐ 7-day ☐ 31-day
- ☐ Dual-speed. Specify: _____ and _____ (must use battery-operated chart drive)
- ☐ Multiple rotation selection (must use battery-operated programmable chart drive)

H. Inking system
- ☐ Bucket-type pens
- ☐ Disposable fiber-tip pens

Other, _____

I. Temperature system
- ☐ Adjustable Mercury bulb with ½-in NPT sliding union adaptable to thermowell (separate socket), 1-in NPT separable socket to fit 1-in threadolet, insertion length of thermowell (excluding threads) _____ inches, length of stainless steel case-compensated capillary _____ ft (if greater than 25 ft, use fully compensated system.)

J. Automatic chart changer
- ☐ Yes. If "yes," check off either battery-operated or conventional spring-wound in part F and check off 24-hour chart rotation in part G. Also, the following specifications apply:
  1. Recorder to be complete with an extension bracket for installation of Mullins automatic chart changer.
  2. Recorder to have slot cut in bottom and plated to accept Mullins automatic chart changer.
- ☐ No

K. Base ratio gear train
- ☐ 1/1 ☐ 10/1 ☐ 100/1 ☐ 1000/1

L. Index rate (dependent on type of primary meter used), ft^3
- ☐ 5 ☐ 10 ☐ 100 ☐ 1000 ☐ Other, _____

M. Index display (uncorrected volume)
- ☐ Digital
- ☐ Dial

III. **Ink Color Standards**

The following ink colors are recommended with scannable blue chart face to satisfy most optical scanner needs.

Differential pen	Red
Static pen	Black
Temperature pen	Green
Cycle pen (for positive displacement or turbine)	Red
Specific gravity	Red

4. The sample cylinder shall be properly purged before any sample is taken.

5. An on-line gas chromatograph should be considered for any station with a flow rate of 50 MMscf/day or above.

D Electronic measurement and control

1. Electronic measurement should be considered for any station. Although complexity, need, economics, and whether the station is new or being upgraded may dictate the final choice, electronic measurement is the best way to handle modern gas measurement business. Figure 3.20 shows a traditional chart-based measurement system. Figure 3.21 depicts a chartless custody transfer system using electronic means. The need for information on a real-time basis, semi-real-time basis, once a day, once a week, or once a month, and the need to remotely control change the scope, magnitude, and cost of a project. Costwise, the low end of electronic flow measurement (EFM) systems, which, in theory replace traditional chart measurement, competes very well with mechanical systems. Various configurations and schemes of data transmission are available. Figures 3.22, 3.23, and 3.24 show basic system philosophy and electronic flow measurement and control concepts. These systems also in the long run reduce operating cost and allow efficient and timely business decisions and operation. With electronic flow measurement, one can have fully automated billing, allocation, and nomination processes.

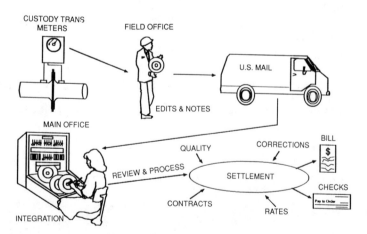

Figure 3.20 Paper chart custody transfer. (*Courtesy: William H. Ryan & American Gas Association*)

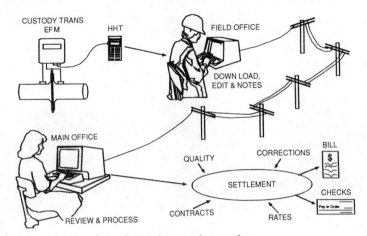

Figure 3.21 Chartless (electronic) custody transfer.

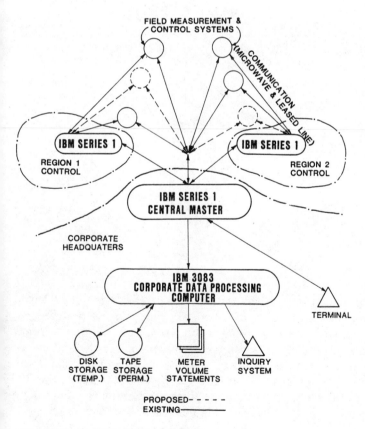

Figure 3.22 Computer network scheme.

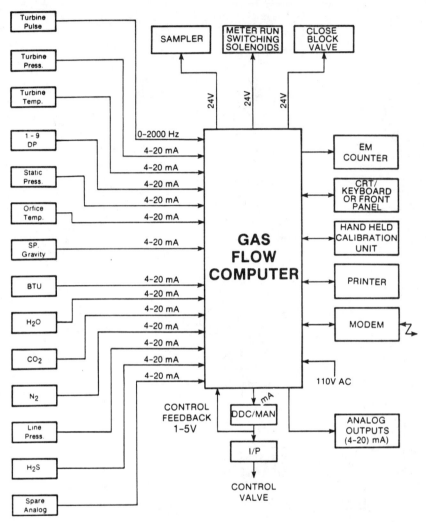

Figure 3.23 Electronic measurement.

2. In general, depending on the type of system (for example, a local printer may or may not be included), all communication and electronic equipment may be outdoors or in a separate building (except for transmitters) in an unclassified area. Some low-cost, low-power systems are intrinsically safe and may be installed in a classified area. In many instances, transducers can be mounted at the flange taps of an orifice run so that gauge lines are eliminated and, as a result, installation cost and possible pulsation effects are reduced.

3. If electronic custody transfer is utilized, flow and temperature recorders are not required. A 31-day backup flow recorder may be desirable

in some instances where installation is remote and no real-time information is available. With electronic custody transfer equipment, a temperature probe should be installed in a thermowell at the tap where traditionally a thermometer bulb is installed.

4. If any regulator or flow control valves are operated by a computer, a pneumatic override system should be installed in the event the computer or associated equipment fails. Meter run switching and proportional-to-flow sampling should be done using an electronic flow computer where practical.

5. All computer inputs should be provided on site and in real time. Minimum inputs required for flow calculation are differential pressure, static pressure, and temperature. Figures 3.22 through 3.24 show examples of different types of systems, network schematic, inputs and outputs, control scheme, etc.

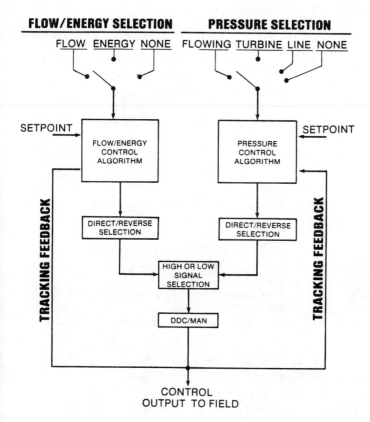

Figure 3.24   Control configuration.

3.3.3   Electrical specifications for classified areas

A. General.   All electrical construction shall be done in accordance with applicable local, state, and federal regulations and must conform to the National Electrical Code (NEC); National Fire Protection Association (NFPA) standard; Classification of Gas Utility Areas for Electrical Installation, Operating Section, American Gas Association; and Title 49, Part 192 of Minimum Federal Safety Standards, U.S. Department of Transportation.

B. Area classification

Class I location	A location in which flammable gases or vapors are, or may be, present in the air in quantities sufficient to produce explosive or ignitable mixtures.
Class I, Division 1 location	A location in which ignitible concentrations of flammable gases are expected to exist under normal operating conditions or in which faulty equipment might simultaneously release flammable gases or vapors and also cause failure of electrical equipment.
Class I, Division 2 location	A location in which flammable gases may be present, but normally are confined within closed systems, or are prevented from accumulating by adequate mechanical ventilation; or the location is adjacent to a Division 1 location from which ignitible concentrations might occasionally be communicated.
Group A	Atmospheres containing acetylene.
Group B	Atmospheres containing hydrogen and other gases.
Group C	Atmospheres containing hydrogen sulfide and other gases or vapors.
Group D	Atmospheres containing butane, gasoline, hexane, methane, natural gas, propane, and most other hydrocarbon gases and vapors encountered in oil and gas production.
Unclassified location	A location not classified as Division 1 or Division 2.
Combustible liquid	A liquid having a flash point at or above 100°F (37.8°C).
Class I	Liquids having flash points below 100°F.
Class II	Liquids having flash points at or above 100°F (37.8°F) and below 140°F (60°C).
Class IIIA	Liquids having flash points at or above 140°F (60°C) and below 200°F (93°C).
Class IIIB	Liquids having flash points at or above 200°F (93°C).

C. Classification of gas utility areas for electrical installations.   Table 3.7 gives electrical classifications for utility areas. Also see Figs. 3.25 to 3.28. (All areas and distances are recommended minimum only and may be exceeded by those responsible for the installation design.)

TABLE 3.7   Electrical Classifications for Utility Areas

Utility area	Class	Group	Division
	Suggested electrical classification		
Natural gas compressor stations			
Compressor building operating floor	1	D	2
Basements, sumps, and trenches:			
Without mechanically induced ventilation	1	D	1
With mechanically induced ventilation	1	D	2
Electrical fixtures attached to compressor building	1	D	2
Auxiliary room, control room, or other room attached to compressor building but separated by a sealed vaportight barrier	Nonhazardous		
Major gas piping area (from fire gate to fire gate):			
Area within 5 ft of any point source of gas pressure relief	1	D	1
Area between 5 and 15 ft of any point source of gas pressure relief	1	D	2
Area within 15 ft of any valve, flange, or screwed connection (outdoors)	1	D	2
Meter or regulator building:			
Bleed gas from controls is vented inside room	1	D	1
Within 10 ft of such room except beyond sealed vaportight barrier	1	D	2
Bleed gas from controls is vented outdoors	1	D	2
General yard area	Nonhazardous		
Gas distribution systems			
All buildings (interior):			
Room contains gas piping; bleed gas from controls is vented inside room	1	D	1
Within 10 ft of such room except beyond sealed vaportight barrier	1	D	2
Room contains gas piping; bleed gas from controls is vented outdoors	1	D	2
Telemetering room, control room, or other room attached to building but separated from hazardous areas by a sealed vaportight barrier	Nonhazardous		
District regulator vault (belowgrade)	1	D	1
City gate stations—major gas piping area:			
Area within 5 ft of any point source of gas pressure relief	1	D	1
Area between 5 and 15 ft of any point source of gas pressure relief	1	D	2
Area within 15 ft of any valve, flange, or screwed connection (outdoors)	1	D	2
Belowgrade locations such as sumps and trenches within Division 1 or Division 2 areas	1	D	1
Residential or commercial meter installations	Nonhazardous		

TABLE 3.7   Electrical Classifications for Utility Areas (*Continued*)

Utility area	Suggested electrical classification		
	Class	Group	Division
Gas transmission systems			
All buildings (interior):			
Room contains gas piping; bleed gas from controls is vented inside room	1	D	1
Within 10 ft of such room except beyond sealed vaportight barrier	1	D	2
Room contains gas piping; bleed gas from controls is vented outdoors	1	D	2
Telemetering room, control room, or other room attached to building but separated from hazardous areas by a sealed vaportight barrier	Nonhazardous		
Regulating and measuring station—major gas piping (fire gate to fire gate):			
Area within 5 ft of any point source of gas pressure relief	1	D	1
Area between 5 and 15 ft of any point source of gas pressure relief	1	D	2
Area within 15 ft of any valve, flange, or screwed connection (outdoors)	1	D	2
Belowgrade locations such as sumps and trenches within Division 1 or Division 2 areas	1	D	1
General yard area	Nonhazardous		

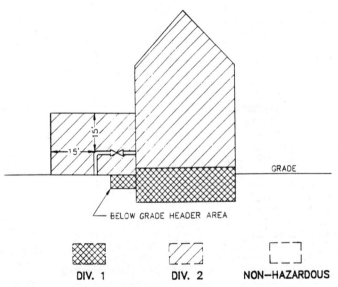

GRADE

BELOW GRADE HEADER AREA

DIV. 1   DIV. 2   NON—HAZARDOUS

Figure 3.25   Compressor building with adequate ventilation.

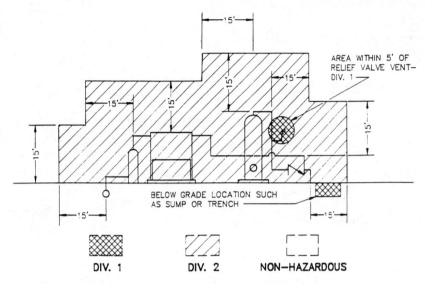

Figure 3.26   Freely ventilated gas piping area.

D. Electrical instruments in classified (hazardous) area at metering and regulating station.   Only those instruments that are rated or designed for hazardous areas (Divisions 1 and 2) can be installed in these areas. In general, these instruments will be either explosionproof or intrinsically safe. All other instruments will be in nonhazardous area. In general, all such equipment shall be at least 15 ft away from the nearest piping. A properly ventilated meter house or enclosure shall be considered Class 1, Division 2. However, rooms or enclosures containing meter and regulator runs, or orifice meter gauges, or natural gas

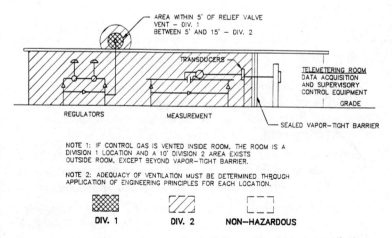

Figure 3.27   Regulating and measuring building with adequate ventilation.

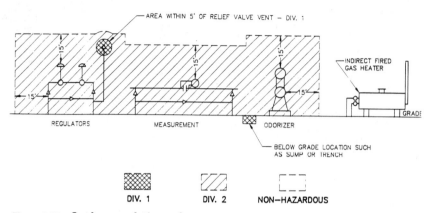

Figure 3.28 Outdoor regulation and measurement area.

odorization units, with control gas being vented inside, shall be considered Division 1 locations. A border of 10 ft of Division 2 area exists around a Division 1 area (Fig. 3.29) unless separated by a sealed vaportight barrier. Area within 5 ft of any relief or blowdown or process equipment vent shall be Division 1.

Telemetering equipment, flow computers, power distribution centers, and open-flame devices such as calorimeters, boilers, water heaters, and emergency generators shall be installed in nonhazardous area and must be at least 15 ft away from the nearest piping or at least 10 ft away from a Division 1 area. However, an intrinsically safe low-power flow computer with low-power transducers may be installed in a classified area, although such equipment may not have explosionproof housing.

In general, classified area installations shall follow Division 1 and Division 2 requirements.

The following construction practices are recommended:

1. *Conduit:* All outside conduit shall be rigid galvanized and, where practical, shall be buried with a minimum of 18-in cover. These buried runs shall be located in such a manner as to avoid interference with existing piping.

2. *Cable:* Power and signal wires shall not be in the same cable or conduit. Instrument wires shall be individually twisted shielded pairs and no. 18 AWG (American Wire Gauge) stranded copper.

3. *Power and telephone service:* Service poles for power and telephone shall be near the fence line. Power and telephone service drops will be installed in buried conduit.

4. *Grounding:* All electrical instruments shall be grounded properly to minimize ground-loop current and to provide a safe path to ground for accidental unsafe voltages.

5. *Location:* Electrical equipment must not be installed within the direct path of discharge from pressure relief valves or blowdown valves.

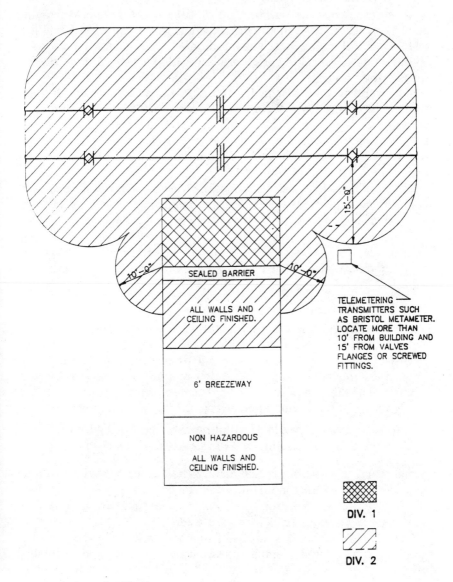

SEALED BARRIER

ALL WALLS AND
CEILING FINISHED.

15'-0"

10'-0"

10'-0"

TELEMETERING
TRANSMITTERS SUCH
AS BRISTOL METAMETER.
LOCATE MORE THAN
10' FROM BUILDING AND
15' FROM VALVES
FLANGES OR SCREWED
FITTINGS.

6' BREEZEWAY

NON HAZARDOUS

ALL WALLS AND
CEILING FINISHED.

DIV. 1

DIV. 2

Figure 3.29 Area classification.

3.3.4 Design specification for electronic equipment building

A. General. This is an example only. Actual specifications will vary from company to company. Controlled environment and a shelter for ease of operation and maintenance may not be a requirement in many situations.

B. Mechanical and construction

General. The basic building may utilize wood or metal framing and, when secured to the foundation, shall withstand wind loading of 50 lb/ft^2 or 150 mi/h winds. The buildings shall be constructed so as to be immune from cracking or otherwise leaking air or water because of weathering or aging.

Floor. Floor shall be constructed of ¾-in exterior grade plywood. Exterior floor shall be covered with a coating of fiberglass and resin type undercoat. Interior floor covering shall be commercial-grade vinyl sheet or tiles edged with perimeter caulking and cemented with waterproof adhesive. Floor shall support any equipment and occupants to a minimum of 100 lb/ft^2.

Walls. The entire rear wall, opposite the door, shall have an insert made of ¾-in plywood for equipment mounting. Openings through walls shall be provided for air conditioner, sleeves, and power service entrance.

Roof. Roof shall be pitched a minimum of 1 in per foot slope to prevent standing water. Roofs shall support a minimum of 50 lb/ft^2.

Finish

1. *Exterior:* Except for door and skid, the entire exterior of building shall be fiberglass or coated metal. If fiberglass buildings are selected, the siding shall be of a blown or molded mixture of polyester resin and chopped fiberglass. Final finish shall be white gelcoat with a 3-mil minimum thickness. If metal buildings are selected, the siding shall be plastic- or vinyl-coated, and all portions shall be highly resistant to rust and corrosion. Fiberglass is preferred.

2. *Interior:* Interior walls and ceiling shall be finished with either ¼-in prefinished paneling or ¼-in prefinished plywood.

Insulation. Walls, floor, and ceiling shall be completely insulated to provide an overall conductance of 0.15 Btu/h at 75°F or 0.14 Btu/ft$^2 \cdot$ °F. The total heat gain shall not exceed 4000 Btu/h. The floor shall have a minimum of 1 in of insulation.

Door. The door shall be of steel frame with one-piece sheet steel or aluminum, insulated, completely weather-stripped, and furnished with stainless steel handle, hinges, and locking device. A heavy-duty closer with open stop pin shall be included. The door is to be hung level with the floor with three hinges on the right as viewed from the outside. It is to be equipped with padlock eyes or hasp and a doorknob or latch of exterior grade on both sides. The door opening shall have an effective rain shield above the door to prevent water accumulation.

Sleeves.   The vendor shall provide three electrical conduit and instrument piping sleeves through the wall centerline, 2 in above the finished floor and centered on the rear wall, opposite the door. Sleeves shall be 1¼-in polyvinyl chloride (PVC) or 1¼-in-PVC-coated rigid steel with a plain end protruding 1 in into building. The outside portion shall have couplings and plugs. If sleeves are steel, the inside portion shall have plastic bushings.

The vendor shall provide one antenna lead-in sleeve through the rear wall, opposite the door, with centerline 3 in from the right wall and 6 in from ceiling. This sleeve shall be PVC and capped on the outside for shipping.

Skid.   An oil-field-type skid shall have a welded frame made from pipe, steel channels, or I beams with sufficient lifting eyes or hook connections for transportation and placement. The frame shall be sandblasted and coated with inorganic zinc and epoxy, or hot-dipped galvanized after fabrication and prior to building erection. The commercial sandblast shall conform to applicable codes (for example, NACE No. 3). The frame shall be constructed so that all support members rest on the foundation.

Foundation.   The foundation will be the responsibility of the specifying company. However, the vendor shall submit the approximate weight of the assembly (building with skid), skid loadings, etc. for proper foundation design.

Dimensions.   All buildings shall consist of a single interior room with interior dimensions as selected below and a ceiling height of 8 ft 0 in (± 1 in). The door shall be 3 ft 0 in wide by 7 ft 0 in high with 1¾ in minimum thickness.

1. 6 ft 0 in (– 0, +1 in) square, flow computer only

2. 6 ft 0 in (– 0, +1 in) × 8 ft 0 in (– 0, +1 in), chromatograph only

3. 8 ft 0 in (– 0, +1 in) × 12 ft 0 in (– 0, +1 in), flow computer and chromatograph

C. Electrical

General.   The vendor shall furnish and install all electrical items. All equipment, devices, conduit, and fittings shall be surface-mounted. All electrical items shall be mounted plumb, level, and square to the building. Electrical work shall strictly conform to the latest edition of the National Electrical Code. Vendor shall furnish and install all electrical items as specified.

Power distribution

1. Circuit breaker box shall be mounted with bottom 5 ft 0 in above finished floor, Square D (or equivalent). Box and interior shall provide eight single-pole spaces, 120/240-V, one-phase, 60-Hz, three-wire, 100-A main circuit breaker, ground bus bar, and surface-mounted cover and door, indoor. The vendor shall provide one electrical power service entrance conduit through the wall from the outside into the circuit breaker box. This conduit shall be 1-in PVC or 1-in-PVC-coated rigid steel with a coupling on the outside and a plastic bushing on the inside. It shall be plugged for shipping. A green no. 6 AWG (moisture-and-heat-resistant thermoplastic type) THWN stranded copper ground wire shall be run from the ground bus in the circuit breaker box to the outside. It shall be bonded to the skid and have 3 ft of slack for connection to a company-installed ground rod electrode.

2. Circuit breakers shall be Square D (or equivalent): three single-pole, 120 Vac, 15 A; two single-pole, 120 Vac, 20 A.

3. Lightning protector shall be Joslyn 1250-33 Surgitron surge arrestor (or equivalent), 120/240 V, one-phase, 60-Hz, three-wire grounded, installed on circuit breaker box and wired to buses.

4. Receptacles shall be mounted 3 ft 0 in above finished floor: two duplex three-wire grounded outlets, 120-V, 60-Hz, with metal covers. Flush-mounted plastic covers will not be acceptable.

Lighting.   Illumination shall be provided by a 48-in rapid-start fluorescent fixture, 120-V, 60-Hz, with two 40-W complete with lamps and a wall switch with metal cover. Flush-mounted plastic covers will not be acceptable. Mount switch 3 ft 6 in above finished floor.

Conduit.   All conduit shall be Electrical Metallic Tubing (EMT) of appropriately sized, ½- or ¾-in only, with necessary fittings, straps, etc. No conduit shall be run along the rear wall unless specified otherwise.

Wire.   All wiring shall be of appropriate AWG, no. 14 or 12, heat-resistant thermoplastic, type THHN solid or stranded copper, with black hot, white neutral, and green ground.

D. Air conditioning.   The vendor shall furnish and install a 120-V, 60-Hz air-conditioning unit with integral thermostat. The air-conditioner opening through the wall shall be made watertight. The unit shall be located 5 ft 6 in above finished floor, and sloped outward to prevent condensation from running into building.

While the unit will be approximately 8000 Btu/h, cooling load calculations can be made for the following conditions:

Temperature outside	110°F dry bulb, 78°F wet bulb
Temperature inside	90°F dry bulb, 72°F wet bulb
Sky	Clear
Time	Solar noon
Date	July 21
Wind velocity	15 mi/h
Latitude	32.5°N

These conditions may vary depending on geographical locations, and representative conditions should be furnished for calculation. The vendor shall furnish the manufacturer's warranty and proof of purchase for the air-conditioning unit.

E. Heating. The vendor shall furnish and install an electric convection-type resistance strip heater with integral line voltage, 120-V, 60-Hz, thermostat. The unit shall be wall-mounted under the air-conditioning unit, and 2 ft 0 in above finished floor.

While the unit will be approximately 1.0 kW, load calculations can be made under the following conditions:

Temperature outside	20°F
Temperature inside	40°F
Sky	Clear
Time	6:00 A.M.
Date	January 21
Wind velocity	25 mi/h
Latitude	32.5°N

Actual representative conditions for specific geographical area should be used. The vendor shall furnish the manufacturer's warranty and proof of purchase for the heater unit.

A combination air-conditioner/heater unit may be used, which would eliminate the wall heater.

4

Gas Quality

4.1 Gas Quality Specifications

Natural gas, by definition, is a mixture of hydrocarbons or hydrocarbons and noncombustible materials, in gaseous state, consisting essentially of methane. However, the amount of each of the components in the gas stream can significantly affect measurement, operation, pipeline efficiency, and above all, customers. Therefore, an acceptable quality specification is assigned in transactions of natural gas. A typical gas quality specification is given below.

Typical specification

1. *Oxygen:* The oxygen content shall not exceed 0.1 percent by volume, and the parties shall make reasonable efforts to maintain the gas free from oxygen.

2. *Hydrogen sulfide:* The hydrogen sulfide content shall not exceed 0.25 grains per 100 ft^3 of gas.

3. *Mercaptans:* The gas shall not contain more than 0.25 grains (of mercaptans) per 100 ft^3 of gas.

4. *Total sulfur:* The total sulfur content, including mercaptans and hydrogen sulfide, shall not exceed 2 grains per 100 ft^3 of gas.

5. *Carbon dioxide:* The carbon dioxide content shall not exceed 2 percent by volume.

6. *Liquids:* The gas shall be free of water and other objectionable liquids at the temperature and pressure at which the gas is delivered, and the gas shall not contain any hydrocarbons which might condense to free liquids in the pipeline under normal pipeline con-

ditions and shall in no event contain water vapor in excess of 7 lb per 1,000,000 ft^3.

7. *Dust, gums, and solid matter:*   The gas shall be commercially free of dust, gums, gum-forming constituents, and other solid matter.

8. *Heating value:*   The gas delivered shall contain a daily, monthly, or yearly average heating content of not less than 975 nor more than 1175 Btu/ft^3 on a dry basis.

9. *Temperature:*   The gas shall not be delivered at a temperature of less than 40°F, and not more than 120°F.

10. *Nitrogen:*   The nitrogen content shall not exceed 3 percent by volume.

11. *Hydrogen:*   The gas shall contain no carbon monoxide, halogens, or unsaturated hydrocarbons, and no more than 400 ppm of hydrogen.

12. *Isopentane+:*   The gas shall not contain more than 0.20 gal isopentane or heavier hydrocarbons per thousand cubic feet.

Effect of various components

Oxygen is naturally not available in natural gas. However, in low-pressure systems, it can get in from the surrounding atmosphere through small leaks. Oxygen can be highly corrosive even at tens of parts per million concentration, depending on circumstances.

Hydrogen sulfide, sulfur, carbon dioxide, and water can form acidic compounds and be highly corrosive to the system.

Nitrogen is inert and harmless. However, it occupies space, thereby reducing pipeline capacity. It consumes horsepower in being moved, and has no energy value. In addition, it can create problems at the burner tip by lifting the flame.

Isopentane and heavier components can drop out as free liquid under normal operating conditions and cause operating problems, besides creating difficulty for customers. Excessive free liquid also reduces the capacity of a pipeline. It also reduces the effectiveness of gas odorization program by absorbing or masking the odorant smell.

High temperature can be detrimental to the coating of pipe. Low temperature can cause freezing problem.

Carbon dioxide and nitrogen also play an important role in determining the supercompressibility factor in flow calculations.

In addition, oxygen, nitrogen, carbon dioxide, hydrogen sulfide, sulfur, water, etc. do not contribute to the heating value of the gas stream. Therefore, they are not part of the desired energy source. In natural gas, we are dealing with energy, and, therefore, these extraneous components are not desirable and must be limited.

4.2   Determination of Gas Composition

General

Gas measurement is not complete without determining gas composition and certain physical characteristics of natural gas. The inerts (CO_2, N_2), specific gravity, etc. enter into actual calculation of volume by the AGA Report No. 3 and AGA Report No. 7 formulas. In addition, with natural gas transactions, what one is dealing with is a package of energy. Therefore, determination of energy (Btu) value is essential. Gas composition also allows one to monitor or comply with gas specifications in a contract. In general, a chromatographic analysis is requested to determine gas composition. However, a typical chromatographic analysis may not provide all the items listed in a quality specifications. Additional analytical instruments are used to determine water vapor, hydrogen sulfide, total sulfur, etc. Nevertheless, chromatographic analysis of natural gas is an integral part of gas measurement business.

Determination of gas composition by the chromatographic method essentially involves two parts:

1. Obtaining a representative sample

2. Analyzing the sample and quantifying the components detected

Sampling

Sampling is done in three different ways:

1. *Spot sampling:*   A sample cylinder is filled with the gas to be analyzed. This, in general, represents the gas at the time of taking the sample. If gas composition does not vary much (i.e., comes from the same well or same formation), a spot sample can be assumed to be representative for a much longer period of time. This is the least expensive way to obtain a sample.

2. *Continuous composite sample:*   In this method, a sample cylinder is filled over an extended period of time (maybe 1 month) by automatically injecting small samples at intervals. Continuous sampling provides a better representation with varying gas composition. Automatic injection of gas samples into the cylinder can be at equal preset intervals of time (proportional to time) or at passage of

equal amounts of gas volume (proportional to flow). In general, if flow rate does not change drastically, a proportional-to-time sample is adequate. With fluctuating flow, a proportional-to-flow sample is recommended.

3. *On-line sampling:* In this method, a gas sample is directly injected from the flowing stream into the analytical instrument or a chromatograph. The result is an almost immediate identification and quantification of the components. A single result like this provides a snapshot of the sample at that instant. However, by repeating this process continuously, one can obtain numerous gas analyses which can then be averaged over a desired period of time. A portable device will provide a spot reading, whereas an on-line device will provide many readings at preset time intervals.

Improper sampling is, in general, the most common cause of bad analyses. Sampling is also a process that is either neglected or not well-understood. For proper sampling one has to understand sampling procedures [GPA (Gas Processors Association) Standard 2166] and sampling techniques, understand the effects of low temperature and pressure loss, and know how to ship sample cylinders, clean sample cylinders, etc.

Sampling techniques

Before a sample can be taken, the empty, clean sample cylinder must be purged thoroughly. Essentially, the GPA-recommended fill-and-empty purge method is used. The minimum number of purges recommended in the fill-and-empty method is given below.

Maximum cylinder pressure, lb/in² gauge	Number of purges
15–30	13
30–60	8
60–90	6
90–150	5
150–500	4
Above 500	3

Some sampling techniques are identical regardless of which type of sampling takes place:

1. The sampling point should be at the 12 o'clock position on the pipe and at a location where the gas is representative. Static areas such as ends of headers, blowdown risers, and the like should be avoided,

as should locations immediately upstream of an orifice where flow distortion from the probe could occur.

2. A sample probe should be used that extends well away from the internal wall of the pipe and into flowing stream. Dirt, liquid, and contaminates such as burned glycol tend to cling to the pipe wall and will crawl into any opening without a probe. This would contaminate an analysis sample.

3. Needle valves should not be used because a pressure and temperature drop could develop over the restricted orifice in the valve. Full, quick-opening ball valves are preferred in the sampling systems.

4. Drip pots, in-line separators, or filters should be avoided except for determining whether liquid is present. The use of drip pots on a line from which a sample is to be analyzed may remove troublesome liquids but may also alter the sample and render it nonrepresentative. Removing liquids may be removing Btu that were in a vapor state in the meter tube.

5. All plumbing should be as short and small as possible; ⅛- or ¼-in stainless-steel tubing is recommended. The shorter and smaller the plumbing, the shorter the time lag and the less the opportunity for absorption and adsorption on the sampling system plumbing walls. Further, the likelihood of contamination from dirt and grease will be less.

6. All piping should slope downward toward the tap so that liquid traps will not form.

7. The sample plumbing must be kept clean. Grease and oil will absorb and desorb heavy hydrocarbons.

Effect of low temperature and pressure loss

Temperature. With continuous samplers, it is not unusual to be sampling when the cylinder temperature is, or will be, below the flowing gas temperature. Also, there is a possibility that the ambient temperature of a cylinder from a remote source may be below the flowing gas source temperature while in transit. If this is the case, condensation can take place in samples containing heavier hydrocarbon components. Although a very small amount of the condensation may be irreversible, almost all of it can be vaporized by preheating prior to directing the sample to the test instrument. The sample cylinders should be heated to approximately 150°F prior to analysis.

Pressure loss. With the continuous sampler, there is virtually no way to guarantee that liquids will not enter the cylinder if aerosols or liquid are present at the sampling source. This is one area where the chromatograph is vastly superior in determining Btu value and specific gravity, because only a small gas sample volume is required. No appreciable pressure drop is encountered when introducing the sample to the analyzer. A significant pressure drop from the sample source to the analysis device can cause fractionation of the heavier liquids inadvertently sampled with the gas. This may occur when a calorimeter consumes the 3 ft^3 of gas required.

Loss of gas pressure from a leaking cylinder being transported from the source to the analysis instrument will normally render the sample nonrepresentative. Therefore, the cylinder connections must always be carefully checked for leaks prior to transport and must be checked for pressure retention prior to directing the sample to the instrument. The lighter ends will normally escape quickly, leaving the heavier, nonrepresentative components in the sample. Also, liquids that appeared in the meter tube and were unavoidably sampled in the continuous sampler could appear as a vapor at the reduced pressure and would not be representative.

Cleaning sample cylinders

Sampling with clean cylinders is very important because residue from a previous sample will erroneously influence the results of a current sample. Preferably, clean cylinders are assigned to the same sampling sources each month to provide additional insurance against contamination.

There are a number of procedures that may be followed to properly clean standard sample cylinders, two of which follow:

1. Flush with volatile solvent and air-dry
   a. Drain all product (gas or liquid) from cylinder.
   b. Hook the sample cylinder to a solvent source.
   c. Open both valves.
   d. Fill cylinder to top with solvent.
   e. Flush solvent through the cylinder (a minimum of 3 minutes).
      Note: If waxy, heavy, sticky, or dirty products are being sampled, time must be increased to ensure cylinders are clean and free of contamination.
   f. Drain the cylinder.
   g. Purge the cylinder with dry air or gas to ensure that no residue remains.
   h. Close valve.
   i. Remove from the cleaning manifold and store properly.
2. Steam-clean and air-dry
   a. Drain all product from cylinder.

b. Hook cylinder to cleaning manifold.

c. Open both valves.

d. Introduce steam into sample cylinder from top to bottom.

e. Flush cylinder with steam for 3 minutes. *Note:* If waxy, heavy, sticky, or dirty products are being sampled, time must be increased to ensure that cylinders are clean and free of contamination.

f. Drain the cylinder.

g. Purge the cylinder with dry air or gas to ensure that no residue remains.

h. Close valve.

i. Remove cylinder from manifold and store properly.

Shipping gas samples in cylinders

Each sample must be in cylinders marked with DOT Specification 3E 1800. Cylinders may be packed one to four per shipping box. Cylinders must be securely packed with crumpled newspaper or other material. The corrugated cardboard box must be 200 lb/in^2 test. The appropriate DOT red flammable-gas label must be affixed to the carton. If United Parcel Service (UPS) is used for shipping sample cylinders, necessary UPS forms and labels must be completed and attached as required by UPS for hazardous materials.

Chromatographic analysis of gas sample

Gas chromatography uses a system for injecting a sample, separating the components, and identifying and measuring the separated components. The elements that make up the system are:

1. A carrier gas for transporting the sample at a constant flow rate. The carrier gas has to be an inert gas and must not react with or change the representative sample in any way.

2. A sampling apparatus to measure and inject the sample into the carrier gas ahead of the component-separating element (column).

3. A chromatographic column containing a stationary phase (packing) consisting of a solid absorbing agent or liquid partitioning agent supported by a solid for separating the sample into individual components because of selective retardation of various components of different molecular weight.

4. A detector for detecting the components transported by the carrier gas.

5. A temperature-controlled chamber housing at least the column and the detector.

6. A device for recording the component peaks that constitute the chromatograph.

In addition to the carrier gas, an extremely high quality, certified calibration gas of known composition must be used for calibrating the chromatograph. The calibration gas should have components generally expected in the unknown sample and its specific gravity should fall within the expected specific gravity range of the unknown.

Operating limits

The following limits are recommended for proper quality control in analyzing a gas sample. If these limits are exceeded, the analysis needs to be investigated. It may require recalibration, a new sample, or other corrective measures. These limits are based on gas chromatograph supplier recommendations and experience.

Heating. Sample bottles are to be heated at least 3 hours at 150°F prior to analysis.

Unnormalized mole percent

For laboratory	98 to 102 percent
For field	95 to 105 percent

Response factor shift

For laboratory	± 5 percent
For field	± 10 percent

Retention time shift. ± 1.5 percent or ±3 seconds, whichever is larger.

Calibration gas pressure. Minimum pressure 20 lb/in^2 gauge.

Carrier gas pressure. Minimum pressure 150 lb/in^2 gauge.

Calibration frequency

Laboratory	Once a day, when operating
Field	No less than once a week

Physical constants

The following table, derived from AGA Report No. 3, 1985, may be used. These constants are given at 14.73 lb/in^2 absolute and 60°F.

Component	Specific gravity (ideal)	Heat value, Btu	GPM†	Mole weight
Methane (C$_1$)	0.5539	1012.0	0	16.04
Ethane (C$_2$)	1.0382	1773.7	26.73	30.07
Propane (C$_3$)	1.5225	2522.1	27.51	44.10
Isobutane (IC$_4$)	2.0068	3260.5	32.70	58.12
N-butane (NC$_4$)	2.0068	3270.1	31.51	58.12
Isopentane (IC$_5$)	2.4911	4011.1	36.59	72.15
N-pentane (NC$_5$)	2.4911	4018.2	36.22	72.15
Neopentane (neoC$_5$)	2.4911	3993.8	38.30	72.15
*Hexane+ (C$_6$+)	3.3127	5288.8	44.62	95.96
Nitrogen (N$_2$)	0.9672	0	0	28.01
Carbon dioxide (CO$_2$)	1.5195	0	0	44.01

*C$_6$+ values provided by gas chromatograph manufacturer or determined from GPA recommendations.
†GPM is liquefiable gallons per 1000 ft^3.

The following constants are from GPA Standard 2145-89 converted to 14.73 lb/in^2 pressure base.

Component	Specific gravity (ideal)	Heat value, Btu	GPM†	Mole weight
Methane	0.5539	1012.3	0	16.04
Ethane	1.0382	1773.7	26.75	30.07
Propane	1.5226	2521.9	27.56	44.10
Isobutane	2.0068	3259.4	32.71	58.12
N-butane	2.0068	3269.8	31.53	58.12
Neopentane	2.4912	3993.8	38.30	72.15
Isopentane	2.4912	4010.2	36.59	72.15
N-pentane	2.4912	4018.2	36.22	72.15
*Hexane +	3.3132	5288.7	44.63	95.97
Nitrogen	0.9672	0	0	28.01
Carbon dioxide	1.5196	0	0	44.01

*C$_6$+ values provided by gas chromatograph manufacturer or determined from GPA recommendations.
† GPM is liquefiable gallons per 1000 ft^3.

4.3 Btu Calculation and Effect of Water Vapor

Water vapor is present to some extent in almost all natural gas. Water vapor affects the measurement of the gas in two ways:

1. *Volume:*   Water vapor occupies volume in the pipeline. A portion of the gas we measure is actually water vapor.

2. *Btu:*   Water vapor has no heating value. Its presence dilutes the natural gas and lowers its heating value (Btu/ft^3).

The presence of water vapor may be corrected for by reducing either the volume or the Btu value. However, both should not be adjusted at the same time.

Water vapor volume

The volume water vapor occupies may be determined by either of two methods, depending on what kind of measurement is made.

1. *Dew point:*   The most commonly used instrument determines the temperature at which the water vapor begins to condense out of the gas. This temperature is called the *dew-point temperature.* If the dew point is known, the partial pressure of the water vapor (Fig. 4.1) may be found in a table of vapor pressures. The ratio of the

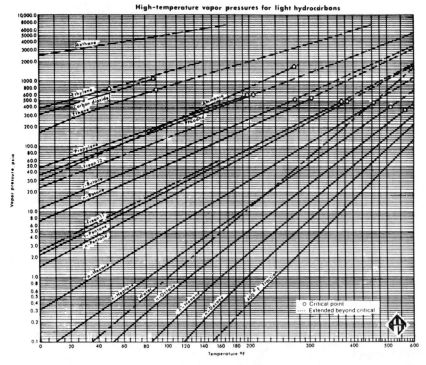

Figure 4.1   High-temperature vapor pressure for light hydrocarbons. (*Courtesy: Gas Processor Suppliers' Association.*)

vapor pressure to the total absolute pressure is proportional to the ratio of the water vapor volume to the total volume:

$$\frac{P_{water}}{P_{total}} = \frac{V_{water}}{V_{total}} = X_{H_2O}$$

where V_{water}/V_{total} = volume fraction of water.

Example   A dew point of 45°F is determined for gas at a pressure of 250 lb/in² gauge. From vapor pressure tables, P_{water} = 0.1475.

$$X_{H_2O} = \frac{P_{water}}{P_{total}} = \frac{0.1475}{250 + 14.7} = 0.0006$$

The gas is 0.06 percent water vapor.

2. *Pounds of water per MMscf:*   Some relatively new electronic instruments are able to determine the water content in pounds of water per million cubic feet (Figure 4.2). This can be converted directly to volume. One pound of water vapor at 14.73 lb/in² absolute and 60°F occupies 21.0181 ft³. Therefore:

$$X_{H_2O} = \frac{lb\ water}{million\ ft^3\ gas} \times 21.0181 \frac{ft^3\ water}{lb\ water}$$

$$= lb/MMscf \times 0.000021$$

Example   The water content is determined to be 23 lb/MMscf.

$$X_{H_2O} = lb\ water \times 0.000021 = 23 \times 0.000021$$

$$= 0.0005$$

The gas is 0.05 percent water vapor.

Dew point can be converted to lb/MMscf, or lb/MMscf can be converted to dew point by using water content curves. It is generally preferred to convert all measurements to lb/MMscf so that contract compliance can be checked.

Volume calculations

The volume of water vapor can be calculated by multiplying water volume fraction X_{H_2O} by the total measured volume:

$$V_{water} = V_{total} \times X_{H_2O}$$

Example   A total of 4,000,000 Mscf was measured during a month. The average water content was 23 lb/MMscf.

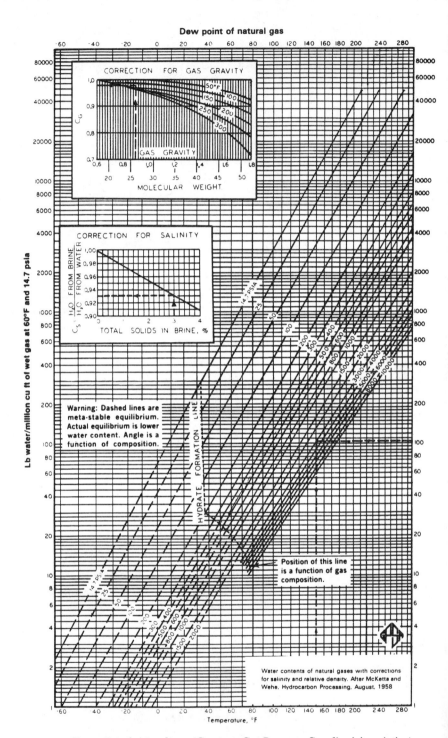

Figure 4.2 Dew point of natural gas. (*Courtesy: Gas Processor Suppliers' Association*)

$$V_{water} = V_{total} \times X_{H_2O} = 4{,}000{,}000 \times (23 \times 0.000021)$$

$$= 2000 \text{ Mscf}$$

Water vapor in excess of contract limit

If a meter volume is being corrected for water in excess of the contract limit (7 lb/MMscf usually), the contract limit should be deducted from the measured water content.

Example   The volume measured is 4,000,000 Mscf. The water content is 23 lb/MMscf. The contract limit is 7 lb/MMscf.

$$\text{Excess water} = 23 - 7 = 16 \text{ lb/MMscf}$$

$$X = 16 \times 0.000021 = 0.0003$$

$$\text{Excess water vapor} = 4{,}000{,}000 \times 0.0003 = 1200 \text{ Mscf}$$

Water vapor in saturated gas

Sometimes gas is purchased which does not meet normal contract specifications. This is usually done because the cost of installing dehydration equipment is prohibitively high (on an offshore platform, for example). In such cases, it is sometimes assumed that the gas is saturated with water vapor. If the gas is saturated, the dew-point temperature is equal to the flowing temperature. If the flowing temperature and pressure are averaged for the chart or billing period, a reasonably accurate determination of the water vapor that was produced can be made.

Example   The volume measured for a month is 4,000,000 Mscf. The average pressure is 800 lb/in^2 absolute, and the average temperature is 105°F.

The water content curve can be used to determine that the average water content for the month was 86 lb/MMscf. This is 79 lb above the normal contract limit of 7 lb/MMscf.

$$X_{H_2O} = 79 \text{ lb/MMscf} \times 0.000021 \text{ MMscf/lb} = 0.0017$$

$$\text{Excess water volume} = 4{,}000{,}000 \text{ Mscf} \times 0.0017 = 6800 \text{ Mscf}$$

If the contract allows, 6800 Mscf may be deducted from the month's production.

Gas saturated at flowing conditions *does not* contain as much water as gas saturated at standard conditions. Gas saturated at 14.73 lb/in^2 absolute and 60°F will hold about 828 lb/MMscf.

Adjusting Btu for water vapor (and pressure base)

The heating value of natural gas is usually determined by a calorimeter or a chromatograph. Most calorimeters record Btu per cubic foot when the gas is at 14.735 lb/in^2 absolute (30 inHg), 60°F, and saturated with water vapor. A chromatograph analysis is used to calculate Btu. Typically, the Btu computed is for a *dry* cubic foot of gas at 14.73 lb/in^2 absolute and 60°F. In either case, the Btu must be adjusted to comply with contract conditions (pressure, temperature, water content).

Btu pressure base

Most contracts stipulate a 14.735 lb/in^2 Btu pressure base. Some make the Btu base the same as the volume base (typically 14.73, 14.65, or 15.025 lb/in^2). The volume and Btu are not necessarily computed at the same pressure base. The measured Btu may be corrected for contract pressure base with the following equation:

$$\text{Contract Btu} = \text{measured Btu} \times \text{PBF}$$

where PBF is the pressure base factor,

$$\text{PBF} = \frac{\text{contract Btu pressure base}}{\text{measured Btu pressure base}}$$

Example

$$\text{Contract Btu} = 1050 \times \frac{15.025}{14.735} = 1050 \times 1.0197 = 1071$$

Btu temperature base

All existing Btu measurements and contracts are at a temperature base of 60°F and no adjustment is required.

Btu moisture base

The Btu per cubic foot may be computed for the gas when it is (1) dry, (2) saturated at base conditions, or (3) at actual flowing conditions. In any case, the adjustment is computed by determining the volume fraction of the water vapor (X_{H_2O}).

Wet to dry conversion

Natural gas which is saturated with water vapor at 14.735 lb/in^2 and 60°F contains about 828 lb water per MMscf.

$$X_{H_2O} \text{ (saturation)} = 828 \times 0.000021 = 0.0174$$

For every cubic foot of gas mixture there is 0.0174 ft^3 of water vapor. This leaves 0.9826 ft^3 of gas. Then

$$\text{Dry Btu} = \frac{\text{wet Btu}}{0.9826} = \text{wet Btu} \times 1.0177$$

where $1/0.9826 = 1.0177$ is the *saturation factor*. Thus,

$$\text{Dry Btu} = \text{wet Btu} \times \text{saturation factor}$$

Example The average Btu recorded by a calorimeter for the month is 1015 Btu/ft^3 (14.735 lb/in^2 absolute, 60°F, wet.) The contract requires the Btu to be on a dry basis.

$$\text{Dry Btu} = 1015 \times 1.0177 = 1033$$

This contract also stipulates a 15.025 lb/in^2 pressure base.

$$\text{PBF} = \frac{15.025}{14.735} = 1.0197$$

$$\text{Contract Btu} = 1015 \times 1.0177 \times 1.0197 = 1053$$

Dry to actual

Again the volume fraction of the water vapor is computed and applied to the measured Btu. The equation shows that part of the dry cubic foot is replaced with water vapor (X_{H_2O}):

$$\text{Actual Btu} = \text{dry BTU} \times (1 - X_{H_2O})$$

where $1 - X_{H_2O}$ is the *moisture factor*. Then

$$\text{Actual Btu} = \text{dry Btu} \times \text{moisture factor}$$

Example A spot sample and moisture test give the following results: Btu = 1050 (14.73 lb/in^2 absolute, 60°F, dry); water content = 50 lb/MMscf; $X_{H_2O} = 50 \times 0.000021 = 0.0011$

$$\text{Moisture factor} = 1 - 0.0011 = 0.9989$$

$$\text{Actual Btu} = 1050 \times 0.9989 = 1049 \qquad (14.73 \text{ lb/in}^2, 60°F, \text{actual})$$

Wet to actual

This calculation is easiest to handle by converting to a dry Btu and then to actual:

$$\text{Actual Btu} = \text{wet Btu} \times \text{saturation factor} \times \text{moisture factor}$$

$$= \text{wet Btu} \times 1.0177 \times (1 - X_{H_2O})$$

Summary of formulas

$$X = \text{volume fraction}$$

$$X_{H_2O} = \frac{P_{\text{water vapor}}}{P_{\text{total}}}$$

$$= \text{lb } H_2O \text{ per MMscf} \times 0.000021$$

$$\text{Water volume} = \text{total volume} \times X_{H_2O}$$

$$\text{Saturation factor} = 1.0177$$

$$\text{Moisture factor} = 1 - X_{H_2O}$$

$$\text{Btu}_{\text{dry}} = \text{Btu}_{\text{wet}} \times \text{saturation factor} \times \text{PBF}$$

$$\text{Btu}_{\text{wet}} = \frac{\text{Btu}_{\text{dry}}}{\text{saturation factor} \times \text{PBF}}$$

$$\text{Btu}_{\text{actual}} = \text{Btu}_{\text{dry}} \times \text{moisture factor} \times \text{PBF}$$

$$\text{BTU}_{\text{actual}} = \text{BTU}_{\text{wet}} \times \text{saturation factor}$$
$$\times \text{moisture factor} \times \text{PBF}$$

$$\text{Pressure base factor (PBF)} = \frac{\text{contract Btu pressure base}}{\text{measured Btu pressure base}}$$

Btu correction factors

Contract conditions, lb/in^2	Convert from 14.735 wet (calorimeter)	Convert from 14.73 dry (chromatograph)
14.65 dry	1.0118	0.9946
14.65 wet	0.9942	0.9773
14.73 dry	1.0174	1.0000
14.73 wet	0.9997	0.9826
14.735 dry	1.0177	1.0003
14.735 wet	1.0000	0.9829
14.90 dry	1.0291	1.0115
14.90 wet	1.0112	0.9939
15.025 dry	1.0377	1.0200
15.025 wet	1.0197	1.0023

Total energy (MMBtu)

In order to determine total Btu at as-delivered conditions, the following formula should be used.

$$\text{MMBtu} = V_1\left(\frac{P_1}{P_2}\right)\text{Btu }_3\left(\frac{P_2}{P_3}\right)F_d\left(1 - \frac{WV_w}{1{,}000{,}000}\right)\left(\frac{1}{1000}\right)$$

where MMBtu = millions of Btu in the gas stream

V_1 = volume of gas stream, Mscf, at P_1 pressure base and 60°F, unadjusted for water content

P_1 = pressure base at which V_1 is measured

P_2 = contract pressure base (e.g., 14.73 lb/in^2 absolute)

P_3 = pressure base at which Btu$_3$ was determined

Btu$_3$ = Number of Btu per cubic foot measured at P_3 pressure base in sample unit volume by calorimeter (normally at 14.735 lb/in^2, saturated)

F_d = 1.0177 (when Btu is measured on a saturated basis) or 1.000 [when Btu is measured on a dry basis (zero water vapor content)]

W = water vapor, lb/MMscf, as delivered

V_w = 21.0181 ft^3 gas per pound water vapor at 14.73 lb/in^2 absolute and 60°F

Contractual Language Related to Gas Measurement

5.1 General

Accuracy of measurement of gas has not only financial implications, it also has serious contractual implications. In addition to revenue considerations, contractual obligations can have far-reaching impact resulting from improper measurement of gas or accounting thereof, thereby resulting in possible litigation. In every purchase, sales, or transportation and exchange contract, general terms and conditions with reference to measurement are clearly spelled out. These general terms and conditions address three typical areas: definitions of terms, gas quality specifications, method of measurement, and measurement equipment.

It is also essential that in the deregulated environment and with open access of pipelines, contracts between parties must address the nomination and allocation process of gas receipts and deliveries. The decontrol of natural gas prices has presented new challenges for producers, consumers, pipelines, and brokers. Now, the consumers have more direct access to the commodity and producers to a larger market. This poses unique challenges to pipeline companies in planning and managing their increasingly complex operations. One major responsibility and challenge facing a pipeline company with multiple transporters is to maintain each individual transporter at a "zero inventory" level while maintaining sufficient line pack, without using it as a storage area to inject into or to withdraw from. In a price-sensitive business, pipeline operation must be flexible enough to respond to short-notice changes. Because of multiple contracts at a single physical metering point, contractual flow of gas may not have any relationship to the physical gas flows. These are complex issues and must be planned for in the contract.

In this new environment, nominations drive the pipeline operation.

After nomination, an operational notice (usually 24 to 48 hours) must be given to accommodate such nominated delivery. However, actual flow is quite often different from what was nominated. An allocation methodology must be established to allocate physical flows to various contracts. Timely and accurate allocation along with timely information flow between producers, pipelines, transporters, and market are essential. In general, three allocation methods are used and agreed upon. They are:

1. Allocation proportional to nomination
2. Allocation to keep firm service whole
3. Allocation by flow-through method where one or more contracts act as balancer by taking the swings

These provisions should be clearly defined in the business part of a contract. These provisions are not dealt with here as part of contractual language. The following pages outline typical language used in the general terms and conditions of a typical agreement that is specific to gas measurement, without using a specific company name. In an actual contract, appropriate company names or names of the parties entering into such agreement must be used. In addition, appropriate contract pressure base (prescribed by various states) and atmospheric pressure must be used.

5.2   Definitions of Terms

The following terms, when used in any agreement incorporating these general terms and conditions, shall have the following meaning. Additional terms may be defined as required by a specific contract.

1. The term *day* shall mean a period of 24 consecutive hours beginning and ending at 7:00 a.m. local time or at such other time as may be mutually agreed to by the parties.
2. The term *month* shall mean a period of 1 calendar month beginning at 7:00 a.m. local time on the first day of such month and ending at 7:00 a.m. on the first day of the next succeeding calendar month.
3. The term *billing month* shall mean the calendar month in which deliveries were made and for which a statement or invoice is being submitted.
4. The term *quarter* shall mean a period of 3 consecutive calendar months beginning at 7:00 a.m. local time on the first day of the

first month and ending at 7:00 a.m. on the first day of the month in which the next succeeding period of 3 calendar months begins.

5. The term *year* shall mean a period of 365 consecutive days beginning and ending at 7:00 a.m. local time, provided that any such year which contains the date of February 29 shall consist of 366 consecutive days.

6. The term *cubic foot of gas,* for the purpose of measurement of the gas delivered hereunder and for all other purposes, is the amount of gas necessary to fill a cubic foot of space when the gas is at an absolute pressure of 14.73 lb/in^2 and at a base temperature of 60°F.

7. The term *Mscf* shall mean 1000 cubic feet of gas at standard conditions.

8. The term *maximum daily quantity* (MDQ) shall mean the maximum daily quantity of gas which the transporter is to receive or deliver at each receipt or delivery point or in the aggregate, in accordance with the terms of the agreement.

9. The term *contract demand quantity* (CDQ) shall mean the maximum volume of gas deliverable by seller to buyer during any one day.

10. The term *pipeline* shall mean this gas pipeline company and shall include the terms *seller* and *transporter* as the context indicates.

11. The term *buyer* shall include any customer that purchases gas from the pipeline, whether or not served pursuant to a filed rate schedule.

12. The term *shipper* shall include any customer that utilizes any of the pipeline's transportation services.

13. The term *customer* shall include both buyers and shippers.

14. The terms *ratably, pro rata,* and similar terms shall be construed to imply only that level of arithmetic precision as is reasonably practicable, taking into consideration all of the circumstances prevailing at the time.

15. The term *British thermal unit* (Btu) shall mean the quantity of heat required to raise the temperature of 1 lb avoirdupois of pure water from 58.5°F to 59.5°F at a constant pressure of 14.73 lb/in^2 absolute.

16. The term *heating value* shall mean the number of Btu produced by the complete combustion, at constant pressure, of the amount of gas which would occupy a volume of 1 ft^3 at a temperature of 60°F, if saturated with water vapor and at a constant pressure of 14.73

lb/in^2 absolute and under standard gravitational force (acceleration of 980.665 cm/s^2) with air of the same temperature and pressure as the gas when the products of combustion are cooled to the initial temperature of the gas and air and when the water formed by combustion is condensed to the liquid state. The gross value so determined shall be corrected from the conditions of testing to that of the actual condition of the gas as delivered (including the conversion from saturated to dry conditions) expressed in Btu/ft^3 and reported at a pressure base of 14.73 lb/in^2 absolute, provided, however, that if the gas as delivered contains 7 lb water vapor or less per 1,000,000 ft^3, such gas shall be assumed to have zero pounds of water per 1,000,000 ft^3 (dry gas).

17. The term *MMBtu* shall mean 1,000,000 Btu.

18. The term *natural gas* shall mean any mixture of hydrocarbons or of hydrocarbons and noncombustible gases, in gaseous state, consisting essentially of methane.

19. The term *balancing units* shall mean the measurement unit used for the purpose of balancing the amount of gas received by the transporter at the transporter's receipt points with the amount of gas delivered by the transporter for the shipper's account at the transporter's delivery points. The balancing unit shall be reported in MMBtu, which shall be determined by multiplying each Mscf of gas so received or delivered by the heating value thereof.

20. The term *fuel and company-used gas allowance* shall be that volumetric portion, expressed as a percent, of all gas received by the transporter into its system which is used in the operation of the transporter's pipeline system and which includes any lost and unaccounted-for gas.

21. The term *equivalent volumes* shall mean the sum of the volumes of gas measured in MMBtu received by the transporter for the account of the shipper at the transporter's receipt points during any given period of time adjusted for plant volume reduction (PVR), separator gas, and the then current fuel and company-used gas allowance, if applicable.

22. The *average atmospheric pressure* shall be assumed to be 14.7 lb/in^2, irrespective of actual elevation or location of the point of delivery above sea level on variations in such atmospheric pressure from time to time, unless specified otherwise.

23. The term *firm rights* shall mean the firm transportation rights in third-party pipeline systems and the firm contract storage rights in any storage company held by the pipeline.

24. The term *transporting pipeline* shall mean the third-party pipelines and the storage company that render firm service on behalf of the pipeline.

25. The term *transporter's system* shall mean the pipeline system of this gas pipeline company, and the third-party systems in which this gas pipeline company holds firm rights to the extent capacity in the third-party systems is committed to this gas pipeline company.

5.3 Quality

Unless otherwise specifically provided in the agreement, all natural gas received or delivered under the terms of the agreement shall be of pipeline quality and shall conform to the following specifications:

1. *Oxygen:* The oxygen content shall not exceed 0.1 percent by volume, and the parties shall make reasonable efforts to maintain the gas free from oxygen.

2. *Hydrogen sulfide:* The hydrogen sulfide content shall not exceed 0.25 grains per 100 ft^3 of gas.

3. *Mercaptans:* The gas shall not contain more than 0.25 grains (of mercaptans) per 100 ft^3 of gas.

4. *Total sulfur:* The total sulfur content, including mercaptans and hydrogen sulfide, shall not exceed 2 grains per 100 ft^3 of gas.

5. *Carbon dioxide:* The carbon dioxide content shall not exceed 2.0 percent by volume.

6. *Liquids:* The gas shall be free of water and other objectionable liquids at the temperature and pressure at which the gas is delivered and the gas shall not contain any hydrocarbons which might condense to free liquids in the pipeline under normal pipeline conditions and shall in no event contain water vapor in excess of 7 lb per 1,000,000 ft^3.

7. *Dust, gums, and solid matter:* The gas shall be commercially free of dust, gums, gum-forming constituents, and other solid matter.

8. *Heating value:* The gas delivered shall contain a daily, monthly, or yearly average heating content of not less than 975 nor more than 1175 Btu/ft^3 on a dry basis.

9. *Temperature:* The gas shall not be delivered at a temperature of less than 40°F and not more than 120°F.

10. *Nitrogen:* The nitrogen content shall not exceed 3 percent by volume.

11. *Hydrogen:*   The gas shall contain no carbon monoxide, halogens, or unsaturated hydrocarbons, and no more than 400 ppm hydrogen.

12. *Isopentane+:*   The gas shall not contain more than 0.20 gal isopentane or heavier hydrocarbons per Mscf.

If, at any time, gas tendered for delivery under the agreement shall fail to conform to any of the quality specifications set forth above, the receiving party may, at its option, refuse to accept delivery pending correction of the deficiency by the delivering party.

If at any time and/or from time to time, the transporter refuses to accept gas hereunder because of the failure of such gas to meet the specifications set forth in this section, the shipper agrees to cease immediately deliveries of such gas upon notification, to include telephonic notification, by the transporter of such refusal. Notwithstanding any other provision of the agreement, if for any reason whatsoever shipper does not cease deliveries of such gas when so notified by the transporter, the transporter and shipper agree that the transporter will not be obligated in any way to transport or to purchase and/or pay for such gas, and the transporter may cause or seek to cause such gas to be shut in at the wellhead. In addition, the shipper will reimburse the transporter for any and/or all costs, fees, and charges of any type that the transporter deems applicable relative to such gas (including but not limited to processing fees, transportation fees, storage fees, administrative costs, and attorney fees). Determination of such charges will be made solely by the transporter. Nothing contained herein shall in any way limit the transporter from seeking and/or utilizing any other remedies that the transporter might have, and does not limit the shipper's obligations to fully indemnify the transporter for any damages or losses that the transporter incurs as a result of such deliveries.

5.4   Method of Measurement and Measurement Equipment

Unit of volume

The unit of volume for measurement of gas for all purposes shall be 1 ft^3 of gas at a base temperature of 60°F and at a pressure of 14.73 lb/in^2 absolute. Where measurement is by orifice meter, all fundamental constants, observations, records, and procedures involved in the determination and/or verification of the quantity and other characteristics of gas delivered hereunder shall, unless otherwise specified herein, be in accordance with the standards prescribed in the 1985 edition of AGA Report No. 3 (ANSI/API 2530), "Orifice Metering of

Natural Gas," with any revisions, amendments, or supplements as may be mutually acceptable to the parties. Measurement by turbine meter, unless specified otherwise, shall be in accordance with AGA Report No. 7 with any revisions, amendments or supplements as may be mutually agreeable to the parties. When positive displacement or turbine meters are used for the measurement of gas, the flowing temperature of the gas shall be assumed to be 60°F, and no correction shall be made for any variation therefrom, provided, however, that the pipeline shall have the option of installing or causing to be installed a recording thermometer should chart measurement be used. If the pipeline exercises this option and installs such a thermometer, correction shall be made for each degree variation in the average flowing temperature for each meter recording. Where measurement is by other than orifice, turbine, or positive displacement meter, standards commonly acceptable in the industry shall be used in the determination of all factors involved in the computation of gas volumes.

Basis

The measurement hereunder shall be corrected for deviation from Boyle's law at the pressures and temperatures under which the gas is measured hereunder by use of the NX-19 formula or table appearing in the manual entitled "PAR Research Project NX-19, Extension of Range of Supercompressibility Tables," AGA Catalog No. 48/PR, published by the AGA in 1963, or as supplemented or amended from time to time, if mutually agreeable to the parties.

Determination of heating value

The heating value of the gas may be determined by a recording calorimeter, or by a chromatograph, continuous gas sampler, or spot gas samples. The arithmetical average of the hourly heating value recorded during periods of flow each day by a recording instrument shall be considered as the heating value of the gas delivered during such day. In the event a continuous gas sampler is installed, the heating value of the composite sample so taken shall be considered as the heating value of the gas delivered during the applicable period of sampling.

If samples are taken, the samples shall be run on the measuring party's calorimeter or chromatograph at another location. The result of a spot sample shall be applied to gas deliveries for the day when the sample is taken and for all following days until a new sample is taken. All heating value determinations made with a chromatograph shall use physical gas constants for gas compounds as outlined in the 1985 edition of AGA Report No. 3 (ANSI/API 2530), "Orifice Metering of Natural Gas," with any subsequent amendments or revisions to which the parties may mutually agree. Heating value shall be determined to the nearest whole Btu.

Determination of flowing temperature

The temperature of the gas flowing through the meter or meters shall be determined by the continuous use of a recording thermometer installed so that it will properly record the temperature of the gas flowing through the meter or meters, should chart measurement be used. The average of the temperature recorded each day shall be used in computing the volumes of gas for that day. Temperature shall be determined to the nearest whole degree Fahrenheit.

Determination of specific gravity

The specific gravity of the gas flowing through the meter or meters may be determined by the use of a recording gravitometer. The average of specific gravity recorded each day shall be used in computing the volume of gas for that day. At the pipeline's election, the specific gravity of the gas flowing through the meter or meters may be determined by a portable gravitometer or chromatograph, by spot gas samples, or by a continuous gas sampler in lieu of a recording gravitometer. In the event spot gas samples are taken or a continuous gas sampler is installed, the samples shall be run on the measuring party's gravitometer or chromatograph at another location. The specific gravity of the composite sample taken from a continuous gas sampler shall be considered as the specific gravity of the gas delivered during the applicable period of sampling. If the specific gravity is determined by a portable gravitometer or chromatograph or from the taking of spot samples, the result shall apply to gas deliveries for the day of the test and for all following days until a new test is taken. All specific gravity determinations made with a chromatograph shall use physical gas constants for gas compounds as outlined in the 1985 edition of AGA Report No. 3 (ANSI/API 2530), "Orifice Metering of Natural Gas," with any subsequent amendments or revisions to which the parties may mutually agree. Specific gravity shall be determined to the nearest one-thousandth.

Equipment

If the pipeline determines that any revision to the measurement facility is needed because of changes in conditions from the original design conditions furnished by the customer, the pipeline shall furnish, own, and install, at the customer's expense and operate and maintain at its own expense any equipment necessary at the metering point. The measuring stations shall be equipped with, as specified by the pipeline, orifice meter runs, orifice meter gauges, recording gauges, or other types of meter or meters of standard make and design, commonly accepted in the industry, so as to accomplish the accurate measurement of gas delivered. At the pipeline's election, a computer, transducers, and other associated sensing devices may be installed to accomplish the accurate measurement of gas delivered hereunder in accordance with AGA Report Nos. 3, 5, 6, and 7, as appropriate, in lieu of mechanical devices with charts. If a computer and associated devices are installed, the values for gross heating value and specific gravity may be entered either manually, not more frequently than once per month, or as real-time data if such data are available. Values for carbon dioxide and nitrogen used in supercompressibility correction determinations shall be entered as real-time data if such data are available or shall be entered manually at intervals mutually agreed upon, but at least once every 6 months.

Calibration and tests of meters

Chromatographs, if used, shall be calibrated by the measuring party not less frequently than once per month against a standard gas sample. All other measuring equipment shall be calibrated and adjusted as necessary by the pipeline but no more than once each month. The other party may, at its option, be present for such calibration and adjustment. The measuring party shall give the other party notice of the time of all tests sufficiently in advance of conducting them that both parties may conveniently have their representatives present. Following any test, any measuring equipment found to be inaccurate to any degree shall be adjusted immediately to measure accurately.

Each party shall have the right, at any time, to challenge the accuracy of any measuring equipment used hereunder and may request additional tests. If, upon testing, the challenged equipment is found to be in error, then it shall be repaired and calibrated. The cost of any such special testing, repair, and calibration shall be borne by the party requiring the special test if the percentage of inaccuracy is found to be 2 percent or less; otherwise, the cost shall be borne by the party operating the challenged measuring equipment.

Access to meters, charts, and records

The other party shall have access at all reasonable times to the measuring equipment and all other instruments used by the measuring party in determining the measurement and quality of the gas delivered under the agreement, but the reading, calibrating, and adjusting thereof shall be done only by employees, agents, or representatives of the measuring party. All records shall be kept on file by the measuring party for a period of 2 years for mutual use of the parties hereto. The measuring party shall, upon request, submit to the other party records from such equipment, subject to return by that party within 30 days after receipt thereof.

Correction of metering errors

If, upon any test, the measuring equipment in the aggregate is found to be inaccurate by more than 2 percent, registration thereof and any payments based on such registration shall be corrected at the rate of such inaccuracy for any period of inaccuracy which is definitely known or agreed upon, but in case the period is not definitely known or agreed upon, then for a period extending back one-half the time elapsed since the last day of calibration.

Failure of meters

If, for any reason, the measuring equipment is out of service or out of repair so that the quantity of gas delivered through such measuring equipment cannot be ascertained or computed from the readings thereof, the quantity of gas so delivered during the out-of-service or out-of-repair period shall be estimated and agreed upon by the parties hereto on the basis of the best available data, using the first of the following methods that is feasible:

1. By using the registration of any duplicate measuring equipment installed by the measuring party if installed and registering correctly

2. By correcting the error if the percentage of error is ascertainable by calibration, test, or mathematical calculation

3. By using the registration of any check measuring equipment of the other party if installed and registering accurately

4. By estimating the quantity of deliveries by deliveries during preceding periods under similar conditions when the measuring equipment was registering accurately

Check measuring equipment

Either party may install, maintain, and operate at its own expense, at or near each pipeline receipt point and/or each pipeline delivery point, such check measuring equipment as desired, provided that such equipment is installed so as not to interfere with the operation of any other measuring equipment.

Whenever any receipt or delivery point is on the premises of the delivering party, the receiving party shall have the right of free use and ingress and egress at all reasonable times for the purpose of installation, operation, repair, or removal of such check measuring equipment.

In the event check measuring equipment is installed by either party, the other party shall have access to the same at all reasonable times, but the reading, calibration, and adjusting thereof and operation and maintenance shall be done only by the party installing the checking equipment.

New measurement techniques

If, at any time during the term of the agreement, a new method or technique is developed with respect to gas measurement, or the determination of the factors used in such gas measurement, such new method or technique may be substituted for the method set forth in this section when, in the opinion of the parties, employing such new method or technique is advisable.

Common Physical Properties of Natural Gas Components

Component	Formula	Molecular weight	Boiling point, °F, at 14.696 lb/in²	Specific gravity	Btu	Liquefiable hydrocarbon, gal/1000 ft³ (GPM)
Methane	CH_4	16.043	-258.73	0.55392	1012.3	16.95
Ethane	C_2H_6	30.070	-127.49	1.03824	1773.7	26.75
Propane	C_3H_8	44.097	-43.75	1.52256	2521.9	27.56
Isobutane	C_4H_{10}	58.123	10.78	2.00684	3259.4	32.71
Normal butane	C_4H_{10}	58.123	31.08	2.00684	3269.8	31.53
Isopentane	C_5H_{12}	72.150	82.12	2.49115	4010.2	36.59
Normal pentane	C_5H_{12}	72.150	96.92	2.49115	4018.2	36.22
Normal hexane	C_6H_{14}	86.177	155.72	2.97547	4766.9	41.13
Normal heptane	C_7H_{16}	100.204	209.16	3.45978	5515.2	46.13
Carbon dioxide	CO_2	44.010	-109.257	1.51955	0.0	
Nitrogen	N_2	28.0134	-320.451	0.96723	0.0	
Oxygen	O_2	31.9988	-297.332	1.10484	0.0	
Air		28.9625	-317.83	1.00000	0.0	
Hydrogen sulfide	H_2S	34.08	-76.50	1.17669	638.6	
Water	H_2O	18.0153	212.00	0.62202	50.4	

At 14.73 lb/in² absolute.
SOURCE: GPA standard 2145-89.

Note: Quite often in chromatographic analysis of natural gas, hexanes and heavier hydrocarbon components are grouped together as hexane+. GPA standard 2261-86 suggests the following values for hexane+ at 14.696 lb/in², 60°F.

	Average molecular weight	Specific gravity	Btu	ft³/gal	gal/1000 ft³
Hexane+	92	3.1765	5065.83	23.2742	42.97

Metric Units and Metric Conversion Tables in the Natural Gas Industry

SI Prefixes

Multiplying factor	Prefix	Symbol
$1\ 000\ 000\ 000\ 000\ 000\ 000 = 10^{18}$	exa	E
$1\ 000\ 000\ 000\ 000\ 000 = 10^{15}$	peta	P
$1\ 000\ 000\ 000\ 000 = 10^{12}$	tera	T
$1\ 000\ 000\ 000 = 10^{9}$	giga	G
$1\ 000\ 000 = 10^{6}$	mega	M
$1\ 000 = 10^{3}$	kilo	k
$100 = 10^{2}$	hecto	h
$10 = 10^{1}$	deca	da
$0.1 = 10^{-1}$	deci	d
$0.01 = 10^{-2}$	centi	c
$0.001 = 10^{-3}$	milli	m
$0.000\ 001 = 10^{-6}$	micro	μ
$0.000\ 000\ 001 = 10^{-9}$	nano	n
$0.000\ 000\ 000\ 001 = 10^{-12}$	pico	p
$0.000\ 000\ 000\ 000\ 001 = 10^{-15}$	femto	f
$0.000\ 000\ 000\ 000\ 000\ 001 = 10^{-18}$	atto	a

Summary of Metric Units of Measurement

Quantity	SI unit symbol	Name	Application
length	mm	millimeter	Precise dimensions
	m	meter	General dimensions
	km	kilometer	Large distances
area	m^2	square meter	General area
	ha	hectare	Area of land
volume	m^3	cubic meter	Volume of gas
	L	liter	
mass	kg	kilogram	
force	N	newton	
pressure	kPa	kilopascal	Gas or fluid pressure
stress	MPa	megapascal	Stress, strength of materials
energy	J	joule	Work, quantity of heat
power	kW	kilowatt	Rate of energy production

Common Conversion Factors for the Natural Gas Industry

Quantity	To convert from	Multiply by	To obtain	Symbol
Volume	cubic feet at 14.73 lb/in^2 absolute and 60°F	0.028 327 84	cubic meters at 101.325 kPa and 15°C	m^3
Energy	Btu$_{60/61}$	1.054 615	kilojoules	kJ
Heating value, saturated	Btu/ft^3 (14.73 lb/in^2 absolute, 60°F)	0.037 887 67	megajoules per cubic meter	MJ/m^3
Heat rate	Btu/h	0.292 948 61	watts	W
Head	ft · lb$_f$/lb$_m$	2.989 067	joules per kilogram	J/kg
Flow volumes	Mcf	28.327 84	cubic meters	m^3
	MMcf	28.327 84	cubic meters	$10^3\ m^3$
	Bcf	28.327 84	cubic meters	$10^6\ m^3$
	Tcf	28.327 84	cubic meters	$10^9\ m^3$
Flow rate	1000 actual ft^3/min	0.471 947 4	cubic meters per second	m^3/s

Metric Conversion Factors

Quantity	To convert from	Multiply by	To obtain	Symbol
Length	inch	25.4	millimeter	mm
	foot	0.304 8	meter	m
	mile	1.609 344	kilometer	km
Area	square foot	0.092 903 04	square meter	m^2
	square yard	0.836 127 4	square meter	m^2
	acre	0.404 685 6	hectare	ha
	square mile	2.589 988	square kilometer	km^2
Volume, solid or liquid	cubic foot	0.028 316 85	cubic meter	m^3
	cubic yard	0.764 555	cubic meter	m^3
	quart (Can.)	1.136 522	liter	L
	gallon (Can.)	4.546 09	liter	L
	gallon (U.S.)	3.785 412	liter	L
	barrel (oil) (42 U.S. gal)	0.158 987 3	cubic meters	m^3
Mass	ounce (avoirdupois)	28.349 523	gram	g
	short ton	0.907 184 74	tonne	t (Mg)
	long ton	1.016 046 908 8	tonne	t (Mg)
	metric ton	1.0	tonne	t (Mg)
Temperature	Fahrenheit	$(°F - 32)5/9$	degree Celsius	°C
	absolute zero			-273.15°C or 0 K
Force	pound force	4.448 222	newton	N
Pressure	pound force per square inch	6.894 757	kilopascal	kPa
	atmosphere	101.325	kilopascal	kPa
	1 inch water column (60°F)	0.248 843	kilopascal	kPa
	1 inch mercury (32°F)	3.386 39	kilopascal	kPa
	1 mm mercury (0°C)	0.133 322	kilopascal	kPa
	1 bar	100.0	kilopascal	kPa
Stress	1000 pound force per square inch (1 ksi)	6.894 757	megapascal	MPa
Energy	British thermal unit (international)	1.055 06	kilojoule	kJ
	kilowatt-hour	3.6	megajoule	MJ
	calorie (international)	4.186 8	joule	J
Torque	1 pound-force-inch (inch-pound force)	0.112 985	newton-meter	N · m
	1 pound-force-foot (foot-pound force)	1.355 818	newton-meter	N · m
Power	horsepower mechanical (550 ft · lb_f/s)	0.745 699 9	kilowatt	kW
	electric	0.746	kilowatt	kW

Natural Gas Volume Conversion Factors for Standard Cubic Foot at Various Reference Conditions to Cubic Meter at Standard Reference Conditions*

Cubic foot reference conditions		Conversion factor
Pressure, lb/in^2 absolute	Temperature, °F	($ft^3 \times$ factor = m^3)
14.4	60	0.027 693 20
14.65	60	0.028 173 99
14.696	60	0.028 262 45
14.7	60	0.028 270 15
14.73†	60	0.028 327 84
14.9	60	0.028 654 78
15.025	60	0.028 895 17

*Standard reference conditions for the cubic meter are specified at a temperature of 15°C and an absolute pressure 101.325 kPa.

†Consumer and Corporate Affairs, Canada, Standards Branch, stipulates that for the purposes of the Gas Inspection Act, 30 inches of mercury at 32°F is equivalent to 14.73 lb/in^2 absolute.

Natural Gas Energy Unit Conversion Factors British Thermal Unit to Joule

Btu	Used by	Definition	Conversion factor
$Btu_{58.5/59.5}$	American Gas Association ASTM (D 1826.64) California Public Utilities Commission	Heat to raise temperature from 58.5°F to 59.5°F	1 054.804
Btu_{60}*	Many export contracts	Heat to raise temperature 59.5°F to 60.5°F	1 054.678
$Btu_{60/61}$	Gas Inspection Act National Energy Board Alberta Petroleum Marketing Commission Alberta Gas Trunk Line	Heat to raise temperature from 60°F to 61°F	1 054.615
Btu_{IT}†	International Steam Tables	1 Btu/lb = 2326 J/kg	1 055.056
Btu_{UK}†	U.K. gas industry	Based on 15°C calorie	1 054.76

*The term 60° Btu has been used to mean either Btu_{60} or $Btu_{60/61}$. The use of this term should be clarified for each case before using the conversion factors.

†These conversion factors have been defined rather than calculated.

Heating Value Conversion Factors (Btu/ft³ to MJ/m³)

For various definitions of British thermal unit and cubic foot to SI standard reference conditions*

Reference conditions for the cubic foot			Conversion factor (Btu/ft³ × factor = MJ/m³)				
Pressure, lb/in² absolute	Temperature, °F	Humidity	$Btu_{60/61}$	$Btu_{58.5/59.5}$	Btu_{60}	Btu_{TT}	Btu_{UK}
14.4	60	saturated†	0.038 771 66	0.038 778 60	0.038 773 97	0.038 787 87	0.038 776 99
14.65	60	saturated	0.038 098 25	0.038 105 08	0.038 100 53	0.038 114 18	0.038 103 49
14.696	60	saturated	0.037 976 88	0.037 983 69	0.037 979 15	0.037 992 76	0.037 982 11
14.7	60	saturated	0.037 966 37	0.037 973 17	0.037 963 64	0.037 982 24	0.037 971 59
14.73	60	saturated	0.037 887 67	0.037 894 46	0.037 889 94	0.037 903 52	0.037 892 88
14.9	60	saturated	0.037 447 84	0.037 454 55	0.037 450 08	0.037 463 50	0.037 452 99
15.025	60	saturated	0.037 130 89	0.037 137 54	0.037 133 11	0.037 146 42	0.037 136 00
14.4	60	dry	0.038 082 09	0.038 088 91	0.038 084 36	0.038 098 01	0.038 087 32
14.65	60	dry	0.037 432 22	0.037 438 93	0.037 434 46	0.037 447 87	0.037 437 37
14.696	60	dry	0.037 315 05	0.037 321 74	0.037 317 28	0.037 330 66	0.037 320 18
14.7	60	dry	0.037 304 90	0.037 311 59	0.037 307 13	0.037 320 50	0.037 310 03
14.73	60	dry	0.037 228 92	0.037 235 60	0.037 231 15	0.037 244 49	0.037 234 04
14.9	60	dry	0.036 804 16	0.036 810 76	0.036 806 36	0.036 819 55	0.036 809 22
15.025	60	dry	0.036 497 79	0.036 504 51	0.036 500 15	0.036 513 23	0.036 502 99

*SI standard reference conditions are specified as a temperature of 15°C and an absolute pressure of 101.325 kPa and free of water vapor.

†Vapor pressure of water taken to be 0.256 11 lb/in²; reference: 1967 ASME Steam Tables.

Glossary of Commonly Used Terms in Gas Measurement

absolute pressure lb/in^2 absolute; sum of gauge pressure plus atmospheric pressure. When calculating volumes of gas, always use absolute pressure.

absolute temperature temperature measured on the absolute temperature scale.

absolute temperature scale scale corresponding to degrees Fahrenheit in which 0°F = 460° absolute. Degrees Fahrenheit + 460 equals degrees absolute.

absolute zero 460 degrees below what we are used to calling *zero* on a Fahrenheit thermometer. At this temperature there is absolutely no heat and no molecular motion.

adjusted volume a volume which has been invoiced but is corrected via adjustment at some time after preparation of the invoice. A volume corrected by applying specific factors.

advanced billing statement computer-generated statements providing month-to-date estimates for each meter station. Not to be considered as final billing volumes. Normally requested on weekly basis.

AGA American Gas Association; organization that developed the equations we use to calculate gas volumes and the physical specifications for meter stations.

AGA Report No. 1 (1930) resulted from the first meeting of AGA. Members developed equations used to calculate gas volumes as well as the physical specifications for meter stations.

AGA Report No. 3 (1956) revision of AGA Report No. 1. AGA members developed more sophisticated factors to correct gas volumes more accurately.

AGA Report No. 3 (1969) amendment to the AGA Report No. 3 established in 1956. The 1969 report includes construction and installation specifications, instructions for calculating volumes from orifice meters, and orifice meter tables.

AID   average inside diameter. In gas measurement, this usually refers to the inside diameter of the meter tube taken 1 in upstream from the upstream face of the orifice plate, i.e., *coefficient AID*.

allocation   a distribution of the production from a number of wells in a particular field or block.

ambient air   the air by which an object is surrounded.

ambient temperature   the temperature of the medium by which an object is surrounded.

American   also known as *Westcott*. American Westcott is a meter manufacturing company which goes by either name.

analyzer   electronic instrument which determines the location of differential and static pressure by relating the position of the two lines to the zero line on a chart.

area   equals length times width when dealing with a two-dimensional flat surface. Examples of units of area are square inches and square feet.

atmospheric pressure   the pressure exerted over the surface of the earth by the weight of the atmosphere. At sea level, this pressure is approximately 14.7 pounds per square inch (lb/in^2).

average pressure   determined by breaking pressure down into equal time intervals, adding all pressures and dividing by the number of time intervals.

average temperature   determined by adding all temperature values at equal time intervals and dividing the total by the number of time intervals.

balance letter   a statement showing gas received vs. gas delivered and any imbalances. Included are current month imbalance, imbalance from prior months, and the sum of these two, or *accumulated,* imbalance.

base coefficient   basic orifice factor used to calculate an orifice meter volume; does *not* include an integrator machine constant.

bellows meter   orifice meter whose differential measuring device consists of two liquid-filled accordionlike bellows. Also known as *dri-flow meter*.

bleed   to drain off liquids or gas, generally slowly, through a valve called a *bleeder*. To *bleed down* or *bleed off* means to slowly release the pressure of a well or of pressurized equipment.

bleed-off valve   also known as *bleeder* valve; used to release pressure of gas within the pipeline to the air.

blind plate   plate which has no orifice and is installed in a locked-on-location (LOL) meter to ensure that no gas passes through that meter or segment of pipeline.

block   a geographic area offshore designated by a number. If developed, it usually has a platform from which more than one gas and/or oil well is drilled and operated (example: Eugene Island Block 32).

blow down   close off a section of pipe and allow the gas to escape into the air for test purposes.

bourdon tube   also known as pressure spring; a flattened metal tube bent in a curve which tends to straighten under pressure.

Boyle's law   states that the volume of any given weight of gas at constant temperature is inversely proportional to the absolute pressure. As pressure increases, volume decreases and as pressure decreases, volume increase.

β ratio   Beta ratio; the ratio of the diameter of the orifice to the diameter of the pipe.

Btu   British thermal unit; a measure of heat energy equal to the amount of heat required to raise the temperature of 1 lb water 1°F when at 60°F.

butane   a light hydrocarbon, consisting of three carbon molecules and eight hydrogen molecules, that is a gas at atmospheric conditions but is easily liquefied under pressure.

bypass   a route by which gas goes around meter instead of through it. Also applies to a route by which gas is diverted around a processing plant.

bypass valve   used to keep the same pressure on both upstream and downstream sides of a meter.

calibrate   to check, adjust, or standardize the graduations of a measuring instrument such as an analyzer.

calorimeter   an apparatus used for measuring the heating value of a gas.

calorimeter roll   strip chart used on a calorimeter showing the continuous recording of the Btu value of gas.

casing head gas   a gas produced in association with oil.

censor   a person, usually holding title of *chart specialist,* responsible for verifying accuracy of orifice chart recordings, marking charts for analyzer, cleaning charts for scanner, deciding whether charts get processed by analyzer or scanner, and making estimates.

Charles' law   expresses the relationship between the volume of a gas and its temperature; the volume varies *directly* with the absolute temperature if the pressure is constant. As temperature increases, so does volume and vice versa.

check meter   a meter used for purposes of dividing the pipeline system into sections or to check another party's measurement. Measures gas delivered from one pipeline section to another or verifies official measurement within acceptable tolerance.

check valve   a valve that permits flow in one direction only.

Christmas tree   aboveground valve assembly at the top of a gas or oil well used to provide pressure control, production rate control, and shut-in service.

chromatograph   instrument which determines the components (molecular makeup) of a gas sample and its Btu value and specific gravity.

city gate   measuring station at which gas is sold to a distribution company from a gas transmission company.

class number   two-digit portion of a meter number that indicates how the gas was handled (i.e., sold—01, transported—11, exchanged—10, company-used—17, etc.).

clock factor   factor calculated to correct orifice chart for either a slow or fast clock. Factor calculated by dividing the time the chart should have run by the time it actually ran.

close-out   A day (usually the fifth working day) of the month at which time all charts and volumes must be completed in the computer system.

county   the geographic area in which a meter is located. In Louisiana, the term *parish* is used instead. See *parish.*

customer number   An identification code assigned to customer contracts.

cycle   a wavelike or humplike marking on the gridded side of a positive displacement or turbine meter chart. Each cycle represents a certain number of Mcf (thousand cubic feet) of gas.

date placed   date on which chart changer places chart on meter. Written on back of chart. The *date placed* for a chart is the *date removed* for the *previous* chart.

date removed   date on which chart changer removes chart from meter. Written on back of chart. The *date removed* for a chart is the *date placed* for the *next* chart.

dead meter   meter which cannot record gas passing because some necessary part is broken.

deadweight tester   designed to give *very accurate* measurement of a wide range of pressures. Used to test the accuracy of pressure gauges and to determine pressures accurately during tests on pipelines.

dehydration   to remove water from a substance. This process is performed on natural gas when there is excessive water vapor or heavier hydrocarbons in vapor state in the gas.

dehydration tower   facility which uses a liquid (glycol) or solid (silica gel) drying agent to remove water vapor from gas. Often these towers are located at compressor stations.

delivery time   the third value printed on an orifice chart by the scanner printer; also known as *count*. The maximum delivery time possible per chart is 800 for older optical scanners and 12,000 for Metrology 12-K scanners.

densitometer   instrument which measures the density of natural gas.

density   the weight of a substance per unit of volume such as the weight of gas per cubic foot.

deviation factor   see *supercompressibility.*

dew point   the temperature at a given pressure at which a liquid begins to form from a gas.

dew-point recorder   a device used by gas transmission companies to determine and record continuously the dew point of a gas.

dial difference   on a positive meter chart, refers to the number of Mcf (thousand cubic feet) of gas that has passed through the meter at line conditions while the chart remained on the meter.

diameter   the length of a straight line through the center of a circle.

diaphragm   flexible wall which separates the compartments of a positive meter.

differential pressure   the difference between two pressures, as across an orifice plate in a gas line. The differential is the difference between the pressure before the gas passes through the orifice and the pressure after it has passed through the orifice.

discharge pipe   the pipe leading *away* from a compressor carrying gas at relatively high pressures.

distribution lines   pipelines which carry gas from transmission lines to customers. Utility companies operate distribution lines.

downstream   refers to location of gas, meter tube, or pipeline *after* the orifice plate. Downstream gas is flowing *away* from the meter.

dri-flow meter   see *bellows meter.*

DRM   direct-reading meter; similar to a house meter in that no charts are used. Gas is measured by reading a series of dials from month to month and using the dial difference.

dry gas   gas that is composed primarily of light hydrocarbon vapors; these vapors do not tend to liquefy under pressure and temperature conditions usually found in the reservoir or at the wellhead. Also, gas after dehydration.

electroscanner   see *scanner.*

entry code   code which identifies type of volume. Entry legend at bottom of meter statement explains meaning of some typical codes.

ethane   a light hydrocarbon, the molecules of which contain two carbons and six hydrogens. Second largest component of natural gas.

exchange   a term used frequently in gas accounting. Refers to an agreement in which two parties literally exchange gas at different metering points. Usually, no money changes hands.

expansion factor   used when gas passes through an orifice plate and the pressure drops and velocity increases, causing the gas to expand and become less dense. The expansion factor adjusts the volume for this expansion.

extend zero   capability of scanner to read the differential as zero when it actually recorded slightly above zero. This is used when hot weather causes mercury in manometer to expand, making chart appear to show gas when there is none.

extension   the square root of the differential pressure multiplied by the square root of the absolute pressure ($\sqrt{hp}$).

farm tap   a tap on a pipeline where gas is sold to a customer (domestic, commercial, or industrial). The sale is made at the pipeline and is metered by the buyer, who furnishes the pipeline company with the volume.

FERC   Federal Energy Regulatory Commission, the federal agency responsible for regulating activities of natural gas pipeline companies and other energy companies.

field   a geographical area in which a number of gas wells produce from a continuous reservoir. A single field may contain several separate reservoirs at varying depths.

flange   a projecting rim or edge at the end of a length of pipe. The flange is drilled with holes for bolting to other flanged fittings.

flange taps   taps that lead from the flanges of a meter tube to an orifice meter for the purpose of measuring differential pressure. These taps are located 1 in upstream and downstream of the orifice plate.

flange up   to join pipes by means of flanges in making final connections on a piping system; to complete any operation (slang).

flash   the vaporized gas that rises from liquefied hydrocarbons.

flowing pressure   the pressure of a gas as it flows through a meter tube. This pressure is measured by a pressure spring and is recorded on a chart.

flowing temperature factor   adjusts volume when gas is flowing at any temperature other than 60°F or 520° absolute (the contract condition).

fluid   any substance that flows. A fluid yields to any force tending to change its shape. Both liquids and gases are fluids.

Foxboro   manufacturer of numerous gas flow measuring devices, including orifice meters, recorders, transducers, and paper charts.

gas   a colorless, tasteless, odorless fluid substance made up primarily of carbon and hydrogen.

gas analysis   a report that comes from the field showing the results of gas analyzed by a chromatograph. This report lists molecular components, specific gravity, and Btu of a sample of gas.

gas bypassed   gas which is routed around a meter when a meter is shut off for testing or repairs. Bypassed gas is not recorded on a chart, but does go to the customer, so must be added to the volume.

gas, dry   natural gas is considered dry if it contains little water. Contracts usually specify that the water content of the gas cannot exceed 7 lb per 1,000,000 ft^3.

gas laws   statements of the relationships between the pressure, temperature, and volume of gas. See Boyle's law and Charles' law.

gas lift   the process of raising, or lifting, gas from a well by means of injecting gas down the well through tubing. Injecting gas increases the pressure in the well, which forces gas back up the well. This process is used when pressure in a well is too low to force gas up to the surface.

gathering lines   pipelines, usually of small diameter, which originate at a gas well and carry gas to transmission lines.

gauge pressure   the force or push of a substance (such as gas) against the walls of its container. Gauge pressure does not include atmospheric pressure and is measured in pounds per square inch gauge (lb/in^2 gauge).

glycol   liquid drying agent used in dehydration towers to remove water vapor from gas.

GPC   gas purchase contract.

GPC number   also known as *parent GPC number*. A contract number assigned to a gas purchase agreement. *Note:* For every GPC number, there exists a meter number.

gravitometer   an apparatus used for determining the specific gravity of gas.

gravity chart   round paper chart used on a gravitometer, on which specific gravity is recorded.

header   a pipe or other fitting which interconnects two or more meter tubes.

heater   apparatus which contains coils around which water can be heated and through which natural gas can be piped to maintain a good flowing temperature of gas.

hydrates   the hydrocarbon and water compound that is formed under reduced temperature and pressure conditions. Hydrates, which resemble snow or ice, often accumulate inside the pipeline and impede gas flow.

hydrocarbons   organic compounds made up of hydrogen and carbon. Natural gas is made up of hydrocarbons.

hydrostatic test   method of testing a section of pipe for leaks. The pipe is filled with water at a certain pressure and shut in. No loss of pressure after a certain time period indicates no leakage.

ID   inside diameter of pipe.

identity file (ID file)   computer record which assigns names and addresses to a coded entity.

inerts   refers to carbon dioxide (CO_2) and nitrogen (N_2) which are found in varying amounts in natural gas but do not burn and therefore add no energy value to gas.

inner trace chart   an orifice chart on which the innermost recording is the static pressure.

integrator   apparatus which can be used to obtain the extension of an orifice chart. The integrator multiplies the differential times the pressure, extracts the square root, and then integrates with time.

interpolate   to calculate a number between two known numbers.

interstate   between states, as in interstate gas flow from Louisiana into Texas.

intrastate   within a state, as in intrastate gas that flows from one part of a state to another.

J volume   volume obtained from an outside company via mail or telephone.

job   a particular process executed by a computer system.

key meter   see *master meter.*

keyed meter   meters which are linked (or *keyed*) in a computer system to one physical meter (master meter) for various information.

L-10 chart (square root chart)   a chart scaled with the numbers 1 through 10 (square roots of 1 through 100); used on orifice meters only.

lb/in² (PSI)   abbreviation for pounds per square inch. In our work, pressure is usually stated as the number of pounds exerted against an area of one square inch.

lb/in² absolute (PSIA)   abbreviation for pounds per square inch *absolute*; equal to the gauge pressure *plus* the pressure of the atmosphere at that point. *Whenever* a volume is calculated, the pressure must be in pounds per square inch absolute!

lb/in² gauge (PSIG)   abbreviation for pounds per square inch gauge; the pressure inside a pipeline caused by the bombardment of gas molecules against the inner walls of the pipe and recorded on a pressure gauge.

leakage section   section of a pipeline system having check meters located at the points where gas enters and exits the system. The entire pipeline system is divided into many leakage sections.

lease number   number on meter statement used only by a certain production company.

line pressure   the pressure of the gas as it travels through the meter tube;

caused by gas molecules bombarding inner walls of pipe. Does *not* include atmospheric pressure.

linear charts   The numerical scale on the chart ranges from 0 to 100 percent. The increments of pressure are at face value, *not* the square root.

load factor report   partially complete computer report which lists total monthly deliveries for all sold, company-used, and transported gas. Gas accounting personnel calculate and post the day of greatest delivery for the month for weekly and monthly charts. Daily chart highest days are calculated by the computer.

location   the actual location of the meter. Sometimes the name of the well is used instead. Found on meter statements.

logical meter number   meter number created for accounting purposes such as for reduction from a volume. *Not* a physical meter.

LOL (Locked on location)   applies to a meter removed from service without being removed physically. A valve is closed, stopping gas flow.

manometer   a U-shaped glass tube containing liquid (usually water or mercury) that is used to measure the pressure of gases or liquids. When pressure is applied, the liquid level in one side of the U rises while the other goes down. Calibrated markings on one side of the U tube permit a pressure reading to be taken, usually in inches of water.

manometer factor   factor which adjusts a volume for the uneven weight of gas pressing down on each side of a U tube in a manometer. Used only on mercury orifice meters.

master file   meter station and meter tube data which are kept as a computer record. This can be combined with current data from the chart processing department to calculate volumes, furnish reports, etc.

master meter   physical meter to which other meters (physical and logical) are keyed or linked in the computer system for various information.

Mcf   1000 ft^3 of gas. Term is commonly used to express a volume of gas.

MDQ   maximum daily quantity; a predetermined volume of gas per day which a company is legally obligated to furnish a customer or receive.

MDQ letter   a letter sent to the sales department stating exact dates and volumes of gas in Mcf when a customer uses more gas than the MDQ (maximum daily quantity) on any given day or days.

mercury meter   orifice meter which uses a mercury-filled U tube (manometer) to measure differential pressure.

meter correction   adjustment that corrects a previously billed volume for errors discovered in the measuring equipment. Can be either a plus or a minus adjustment.

meter number   an identification number assigned to each meter. It may be composed of the following code numbers: company (2 digits), district (2 digits), class (2 digits), location (3 digits), tube—also referred to as series or run—(1 digit).

meter statement   computer-generated statement on which is printed all completed volumes for each meter number. These statements come out once a month after billing and list daily, weekly, and monthly volumes including factors used in the volume calculations.

meter station   the location at which a meter or meters are physically placed.

meter tube   the portion of pipe which brings gas through the meter station and is physically tapped to the meter.

methane   a light, flammable gaseous hydrocarbon consisting of one carbon and four hydrogens (CH_4). It is the main component of natural gas.

MMBtu   1,000,000 Btu, the unit of measure used to determine amount of gas on a Btu (heating value) basis.

molecule   the smallest part of a substance that can exist on its own. It usually consists of a group of atoms that are either different (CH_4) or alike (H_2). CH_4 is a molecule of methane.

multimeter table cross-reference   computer-generated list of all master meter numbers and those meter numbers that are keyed to them. It also indicates which master meter to refer to for temperature, gravity, Btu, and inert data. Any percentage of volume keyed from the master meter to the keyed meter is also listed.

natural gas   a substance which is usually found in a gaseous state under normal conditions. Consists of molecules which are not all alike, because natural gas is a *mixture* of gases. These gases are all called *hydrocarbons* because they are made up of *hydrogen + carbon* atoms. The lightest hydrocarbon, *methane,* has only one carbon atom and four hydrogens and is the major constituent of natural gas.

no delivery   a condition in which no gas passes through a meter. The condition gives a recording of 0 on an orifice chart, and there will be no cycles on a positive displacement or turbine meter.

noncoincidental max day   used when calculating a load factor report. It is the day of the month or year having the largest gas delivery. Each meter station within a billing area will have its own noncoincidental max day.

octane   a heavy hydrocarbon having 8 carbon + 18 hydrogen atoms; a liquid at atmospheric conditions. Used in making gasoline.

odorant   a chemical that is added to natural gas so that the presence of the gas can be detected by smell. Used in populated areas of a system as a warning signal in the case of gas leakage. Odorant has to be injected to comply with governmental requirements (for example, those of the DOT).

off-system volume   also called *J volume*. Volume which has been measured and calculated by an outside company.

on-system volume   volume which has been measured into or out of a pipeline.

orifice   round hole in the center of an orifice plate. See *orifice plate*.

orifice chart   chart on which an orifice meter records the differential pressure and static pressure.

orifice meter   an instrument used to measure the flow of gas through a pipe. This device measures and records the pressure differential created by the passage of gas through an orifice in the plate that is placed in the pipe. Meter also measures and records the pressure of gas in the line or static pressure.

orifice plate   flat meter disk or plate with a hole in the center. Installed in a meter tube to partially restrict the flow of gas. Its purpose is to create a pressure drop which can be measured and used to determine the volume of the gas passing.

orifice pressure drop   the difference in pressures (pressure differential) that occurs as gas passes across an orifice plate.

orifice thermal expansion factor   correction to volume for change in orifice size. The diameter of the orifice in the center of the orifice plate increases and decreases slightly depending on gas temperature.

outer trace chart   orifice chart on which the outermost recording is the static pressure.

overrange   beyond the range of the chart; exceeding the highest value on the graph.

parent GPC number   the gas purchase contract number used by the chart processing department. See *GPC number*.

parish   in Louisiana, the geographic area in which a meter is located. See *county*.

PC factor   Proportional correction factor, also called *clock factor*. Factor that corrects the dial difference for a fast or slow clock. Determined by dividing the time that should have been recorded on the chart by the time that actually was recorded.

peak day report   report calculated yearly by the gas accounting section to determine the 3 days of highest delivery and/or the 3 *consecutive* days of highest delivery for the year.

physical meter   a meter which actually exists and measures gas somewhere along the pipeline system.

pi (π)   a mathematical constant which equals 3.1416. The ratio of a circle's diameter to its circumference.

pig   a device used to clean the interior walls of a pipeline. Gas pressure behind the pig forces it through the pipeline.

pipe taps   a connection to a pipeline for an orifice meter. The upstream tap is placed 2½ pipe diameters from the orifice plate and the downstream tap placed 8 pipe diameters from the orifice plate.

placed date   the date on which the chart changer places a chart on the meter. It is written on the back of the chart.

platform   an immobile offshore structure constructed on pilings from which wells are drilled and produced.

positive chart   round paper chart on which a positive meter records the dial difference, in cycles, and the static pressure.

positive meter   a device that measures volume of gas by filling and emptying chambers of a specific volume; also called *positive displacement meter*. Meter actually counts and records the cubic feet of gas which pass through the meter.

pressure   the force per unit of area that is exerted on a surface (as that exerted against the inner walls of the pipe by gas molecules). Usually expressed in pounds per square inch (lb/in^2).

pressure base   the number of pounds per square inch absolute (lb/in^2 absolute) used as a standard for expressing gas volumes. A commonly used pressure base is 14.73 lb/in^2 absolute.

pressure delete   a way of operating an electroscanner that can be selected by the operator and causes the scanner *not* to accumulate counts for static pressure.

pressure drop   decrease of pressure which is caused by various factors such as the partial restriction of gas flow by an orifice plate in the line, regulators, or friction between flowing gas and interior walls of pipe.

pressure factor   factor used to convert the volume at the actual flowing pressure (250 lb/in^2 absolute) to a volume at the company's pressure base (14.73 lb/in^2 absolute); also called *pressure multiplier*.

$$\text{Pressure factor} = \frac{\text{flowing pressure (gauge)} + \text{atmospheric pressure}}{\text{pressure base}}$$

pressure in percent of scale   pressure expressed as a percentage rather than as a number of pounds per square inch.

pressure spring   piece of hollow, coiled metal tubing used in orifice meters to record static pressure.

proof list   a computer-generated report received daily which lists all chart data and J volumes that were transmitted the previous day.

propane   hydrocarbon made up of 4 carbon and 10 hydrogen atoms. Is a gas at atmospheric conditions and is a minor constituent of natural gas. Easily liquefiable under pressure.

purge   to thoroughly clean or empty a pipeline (or a piece of equipment).

PVR   plant volume reduction. When gas goes through a *plant* it undergoes conditions of temperature and pressure which cause the heavy hydrocarbons (octanes, hexanes) to separate from the lighter hydrocarbons (methane, ethane). The valuable heavy hydrocarbons are drained off and kept by the plant. The *volume* of gas we started out with is less, or *reduced,* as a result.

radius   half the length of the diameter of a circle.

Ranarex   portable instrument which can be connected to a gas supply (meter tube) to determine its specific gravity. This is called a *spot gravity measurement.*

rangeability   range of volume from lowest to highest which a meter can accurately measure.

Rankine   another term for the absolute temperature scale. See *absolute temperature scale.*

recording Btu   Btu value that is recorded on a calorimeter.

recording gravity   specific gravity which is recorded on a chart by a gravitometer (see *gravitometer*).

recording index   series of dials associated with a positive displacement or turbine meter which displays the number of cubic feet which has passed through the meter. Chart changers read and record this index reading on charts in Mcf.

recording temperature   the temperature recorded on a chart by a recording thermometer.

recording thermometer   thermometer which is inserted in a small well or casing in the pipeline and is connected by a thin mercury-filled tube to a temperature meter. As temperature changes, the mercury in the tube expands or contracts, causing a measuring spring to position a pen on a chart.

regulator   pressure-limiting device used along a pipeline system to reduce pressure to a set value and maintain that constant pressure. See *pressure drop.*

rejected volumes   volumes that are rejected by the sales or purchase system interface.

removed date   the date when the chart changer removed a chart from the meter. The chart changer writes this date on the back of the chart.

resend   to rekeypunch the data from a chart or transmittal to correct that data in the computer system.

residue gas   the gas that is left over after the heavy hydrocarbons have been removed.

Reynolds number factor   ratio of the rate of flow to the viscosity of a gas; always equals 1 or slightly more.

rotor   the portion of a turbine meter which is turned by the flow of gas.

run number the last digit of a meter number which indicates how many tubes are at that meter station; also known as *tube* or *series* number. *Run number* is also used to refer to the sequence in which reports are generated by the computer.

rural service a sale to a gas distribution company which provides gas to a rural community.

saturated Btu the heating value of a gas before water has been removed from it.

saturated gas gas that has not had the water removed from it or has been totally saturated (as in a calorimeter); also called *wet gas*.

scanner an electronic instrument that relates the differential pressure and static pressure lines to the zero line on an orifice chart; commonly called *electroscanner*. The scanner uses this data to determine the extension, which it integrates with time, pressure in percent of scale, and delivery time.

scan value the number of scanner counts registered by the scanner or analyzer for an orifice chart which represents the extension of the chart, integrated with time.

scrubber a vessel through which gas is passed to remove dirt and other foreign materials.

senior fitting device placed directly on a meter tube to hold an orifice plate in place within the line. It allows inspection and changing of the orifice plate without interrupting the flow of gas.

separator a piece of production equipment used to separate the liquid components of the well stream from the gaseous components. Separation of the oil and gas is accomplished primarily by gravity, the heavier liquid hydrocarbons falling to the bottom and the lighter gaseous hydrocarbons rising to the top. The gas then leaves from the top of the separator, and the oil leaves from the bottom.

sequence number a numerical code which is used to allow more than one volume per date or to indicate a prior-month volume. 0 is used to indicate that a chart is for the current month's business or that only one chart was placed on the meter on the date.

series number see *run number*.

shut in to close the valves on a well so that it stops producing. Also used to refer to a section of pipeline that has been closed off.

shut-in test a test of a section of pipeline for leakage. The volume of a section of pipe is measured, the line is shut-in for a period of time, and, at the end of that time, the volume is measured again. No loss of volume indicates that the line has no leakage.

specific gravity ratio of the weight of a cubic foot of natural gas to the weight of a cubic foot of dry air, under the same pressure and temperature conditions. Air is assigned a reference of 1.0, therefore, if the specific gravity of gas is 0.6, this means the gas weighs six-tenths as much as air.

specific gravity factor   a factor applied whenever gas has a specific gravity other than 1.0; necessary because the American Gas Association equations for calculating volumes are based on gas having a specific gravity of 1.

$$\text{Specific gravity factor} = \sqrt{\frac{\text{old gravity}}{\text{new gravity}}}$$

spot Btu   Btu value determined by the gas passing through the meter tube at the time of the test only; can be obtained from spot sample on chromatograph.

spot gravity   specific gravity of gas determined by using a Ranarex; represents the gravity of the gas passing through the meter tube at the time of the test only; can be obtained from spot sample on chromatograph.

spot sample   a sample of gas which is drawn out of the pipeline into a cylinder, called a *sample bottle,* for the purpose of determining the Btu value and/or specific gravity of that gas.

square root   a divisor of a number that when squared gives the number.

square root chart   an orifice chart having a numerical scale of the numbers 1 through 10, which represent the *square roots* of numbers 1 through 100; also known as an *L-10 chart* or *direct-reading chart.*

standard cubic foot   as defined by the AGA, the amount of gas necessary to fill a cubic foot of space at a pressure of 14.73 lb/in^2 absolute and a temperature of 60°F; abbreviated scf.

state standard pressure base (SSPB)   the pressure base at which *purchase* volumes must be calculated in a given state. For example, SSPB is 14.65 lb/in^2 absolute in Texas, 15.025 lb/in^2 absolute in Louisiana.

statement   see *meter statement.*

static pressure   the pressure against the inside walls of a pipeline resulting from the motion of the gas molecules within.

station number   the portion of a meter number which identifies the meter but does *not* refer to geographic location.

storage gas   gas that is stored in an underground reservoir for later use. Gas is measured as it enters and leaves these storage areas.

straightening vane   a bundle of tubing which is installed in a meter tube upstream of the orifice plate to straighten the swirling gas stream. This helps create a more accurate differential pressure across the orifice.

strip chart   the type of chart used on a calorimeter. A length of calibrated paper on a roll which records either for the length of the roll or for the length of time it is on the calorimeter.

suction pipe   the section of pipeline that carries a product out of a tank to the suction side of the pumps.

supercompressibility   the deviation of volume from that predicted by Boyle's law. When changing from the measured pressure (e.g., 200 lb/in^2 absolute) to the pressure base (e.g., 14.73 lb/in^2 absolute), a gas actually expands a little more than Boyle's law indicates.

supercompressibility factor   adjusts the calculation of gas volumes to take into account the deviation from Boyle's law; also called *deviation factor* or F_{pv}.

temperature   is a measure of hotness or coldness; a partial indication of the amount of heat in a substance.

temperature base   the temperature at which the cubic foot is the unit of measurement according to the contract. For our purposes, this temperature is 60°F or 520° absolute and is used as a standard for expressing all gas volumes.

temperature chart   round paper chart on which the temperature of flowing gas is recorded.

temperature factor   factor used to convert a gas volume from its actual temperature to the temperature base (520° absolute).

$$\text{Temperature factor} = \sqrt{\frac{\text{temperature base, absolute}}{\text{average flowing temperature, absolute}}}$$

$$= \sqrt{\frac{520°}{°F + 460°}}$$

temperature/gravity clerk   clerical position in a chart processing department responsible for reading and averaging temperature and gravity charts and assisting the censor.

transmission lines   the part of a pipeline system which carries gas from gathering lines (near the well) to distribution lines (which supply homes, etc.), industrial customers, or other gas transmission companies.

tube   a section of pipe (upstream and downstream) that is flanged at each end and connects an orifice meter to the pipeline.

tube number   a term used interchangeably with run number and series number. This number at the end of the meter number indicates how many tubes are at that station. See *run number*.

turbine meter   an instrument which uses flowing gas as a source of energy to turn a rotor. This turning rotor causes a series of dials to record the number of cubic feet which have passed. The volume passing through the meter is directly proportional to the speed of the flowing gas.

turbine meter chart   a round paper chart on which both volume and static pressure are recorded. *Note:* positive displacement and turbine meters both use the same type of chart.

turbulence   swirling or crosscurrents in the gas stream which can cause inaccurate measurement of volume. See *straightening vane*.

unaccounted-for gas   the difference between the gas delivered and gas received within a given leakage section; it is either a loss or gain.

unaccounted-for gas report   monthly (or yearly) report which itemizes all gas volumes delivered and received by pipeline leakage sections. The unaccounted-for gas is shown as a loss or gain. Used to reconcile the difference between gas received and delivered and to determine reasons for loss or gain of gas. Also gives "system unaccounted-for."

uniform scale chart   a type of orifice meter chart on which the differential pressure scale is expressed in inches of water and static pressure is in pounds.

upstream   refers to position of gas, meter tube, or pipeline *before* the orifice plate. Upstream gas is flowing *toward* the meter.

valve   a device to control the rate of flow in a line, to open or shut off a line completely, or to serve as an automatic or semiautomatic safety device.

vapor   any substance in its gaseous state that is capable of being liquefied by compression and/or cooling.

velocity   rate of flow (speed) of gas.

viscosity   the resistance of a fluid to flow because of the effect of cohesion. See *Reynolds number factor*.

volume   a three-dimensional measurement of the amount of space enclosed in a vessel. For our purposes, an amount of gas expressed in cubic feet or thousand cubic feet.

WACOG   weighted average cost of gas. A monthly report which takes into account many factors including the volume of unaccounted-for gas to determine how much to charge certain industrial customers for the gas.

water manometer   a glass U-shaped tube containing water that is used to test differential pressure on an orifice meter.

well   a hole drilled deep into the earth which allows gas and/or oil to escape to the surface.

wellhead   the equipment used to maintain surface control of a well. It is formed of the casing head and tubing head. The *Christmas tree* is affixed to the top of the tubing head.

well owner   individual or company with which pipeline or end user company has a contract.

Westcott   also known as *American*. A meter manufacturing company. See *American*.

wet   term refers to gas or Btu of gas that contains a high percent of water. See *saturated Btu* and *saturated gas*.

zeroing a meter   procedure done by chart changer or measurement technician to make sure the differential pressure pen of an orifice meter is in calibration. With the pressure at zero, the pen should also read zero.

Limited Glossary of Regulatory Words and Terms

annualization   to adjust to a full-year basis any item included in base period data for less than a full year.

base period   the most recently available 12 consecutive months of actual experience.

certificate   full-term "certificate of public convenience and necessity." A natural gas company, as defined, is required to file with the *Commission* an application for a certificate setting forth extensive information required to advise the Commission fully concerning the operation, sales, service, reserves, construction, extension, and acquisition for which the certificate is required, or for the abandonment of property or service for which permission and approval is required. In the case of a natural gas pipeline company, this means all facilities, except those used solely for gas gathering, which are used in interstate commerce. A certificate is required prior to the beginning of construction, extension, or abandonment of facilities.

Commission   either Federal Power Commission (FPC) or Federal Energy Regulatory Commission (FERC).*

commodity component   that part of the cost of service which must be recovered through use of a *commodity rate,* i.e., a rate per Mcf for every Mcf of gas sold. Revenue from a commodity rate varies as the gas available for sale varies.

cost of service   the total number of dollars required to return to the company all its costs, i.e., operating and maintenance expenses, cost of gas, and other necessary costs such as taxes (including income taxes), depreciation, depletion and amortization of property costs, and over and above all this, a fair return on *rate* base.

*Effective October 1, 1977, the FPC was abolished and its powers and duties were transferred to the FERC, a division of the newly created cabinet level Department of Energy (DOE).

cost zones   geographic areas of the company's operations established for the purpose of accumulating certain costs to facilitate a fair distribution of such costs among all customers.

curtailment   the reduction in the volume of gas delivered to a customer below such customer's needs or contract requirements.

demand component   that part of the total cost of service which must be recovered through use of a *demand rate,* i.e., a rate per Mcf for each Mcf of gas representing the customer's demand on the company's system—usually the highest volume delivered in any day during the past year. This demand volume is used for billing until (1) a higher volume is delivered or (2) 12 months have passed, whereupon a new demand billing volume will be established. Generally, demand revenue will not fluctuate with changes in the annual volume of gas sold.

depreciation   return *of* investment through inclusion in cost of service and rates of a pro rata part of the cost of property, calculated to spread the total cost over a certain period of time or number of units that measure the useful life of the investment. Depreciation is to reimburse the company "…the loss in service value not restored by current maintenance, incurred in connection with the consumption or prospective retirement of gas plant in the course of service from causes which are known to be in current operation and against which the utility is not protected by insurance. Among the causes to be given consideration are wear and tear, decay, action of the elements, inadequacy, obsolescence, changes in the art, changes in demand and requirements of public authorities, and, in the case of natural gas companies, the exhaustion of natural resources." (Quotation is from FERC *Uniform System of Accounts.*)

DOE   Department of Energy.

end use   the actual purpose for which gas is used by the ultimate consumer to whom it is delivered.

equity   that part of a business enterprise owned by the stockholders. Usually represented in the financial statements (balance sheet) of a company as the value of outstanding common and preferred stock, retained earnings and any additional capital. (The remainder of the value of the company is usually represented by outstanding long-term debt, amounts currently owed to others, and various deferred items.)

fair   a subjective term meaning different things to different people, as in "fair return on investment." It was, no doubt, intended by those who first used it in connection with rate regulation as a level of return, or profit, which everyone would concede was "fair." (In practice it means whatever the Commission and/ or the courts say it means.)

FERC   Federal Energy Regulatory Commission.

fixed cost   certain costs, such as return, taxes, and depreciation, which do not vary as the volume of gas sold or transported varies.

FPC   Federal Power Commission.

incremental pricing   the separate pricing for sale of specific volumes of purchased gas at the cost of that particular gas, instead of adding such cost to the cost of all other gas purchased and developing a total system average unit cost of gas.

interstate   with respect to natural gas companies, the transporting and sale of gas for resale across state lines.

intrastate   with respect to natural gas companies, the transporting and sale of gas totally within the boundaries of a state, i.e., not crossing a state line.

jurisdictional   that part of a natural gas company's business which is subject to control and regulation by the Federal Energy Regulatory Commission. Generally, the Commission has *rate* jurisdiction over transportation and sales of gas for resale in interstate commerce, i.e., across state lines. Generally, the Commission has *certificate* jurisdiction over those facilities (except purely gathering) used to transport gas across state lines in interstate commerce.

just and reasonable   another subjective term (see *fair*) with no absolute definition against which compliance can be measured. (In practice it means whatever the Commission and/or the courts say it means.)

line pack   the volume of gas required to fill a pipeline before deliveries can be made. In a rate case *line pack* represents the total cost of such gas, at its average cost per Mcf, which has been capitalized and included in the working capital part of the rate base.

LNG   liquefied natural gas; natural gas which has been converted to a liquid for transportation, usually by oceangoing vessels or barges. It must be converted back to a gaseous form to be used.

locked-in period   that period of time during which a rate or rates are in use, between the date such rates became effective and the effective date of superseding rates. Generally, a different approach is required to determine a cost of service for a locked-in period than for a future period in a rate filing.

LPG   liquid petroleum gas, the most common forms of which are butane and propane, used in many homes in rural areas. LPG, with change in pressure and temperature, changes to a gaseous form and is used like natural gas.

Mcf   1000 cubic feet; a unit used in the measurement of natural gas.

MSAC   maximum surcharge absorption capability. According to the Natural Gas Policy Act, the monthly total of the dollar difference for all low-priority users served between the dollars produced for each such customer by multiplying the customer's volume by the present rate and by the fuel oil equivalent rate. Thus, MSAC is the difference, in dollars, that the user would pay under existing contracts and with respect to an equivalent amount of alternative fuel.

natural gas company   a defined term under the Natural Gas Act meaning a company engaged in the transportation and sale for resale of natural gas in interstate commerce, i.e., across state lines.

NGPA   Natural Gas Policy Act, enacted in November 1978.

nonjurisdictional   generally used to denote business or companies not subject to control and regulation by the Federal Energy Regulatory Commission. *Note:* Much of the natural gas business not subject to control and regulation by the Commission and, therefore, referred to as *nonjurisdictional* is subject to regulation by state regulatory agencies.

normalization   adjustment of base period data to include the annual effect of changes in revenues and costs (including plant) which are known and are measurable with reasonable accuracy at the time of the filing and which will become effective by the end of the test period.

original cost   The actual cost of land, building, pipelines and other plant items "to the person first devoting it to public service," distinguished from the cost to a subsequent owner acquiring such property after it is already "devoted to public service." (The quotations are from the FERC *Uniform System of Accounts.*)

peak day   The one day (24 hours) of maximum system deliveries of gas, usually a day of actual experience occurring during the base period or a day estimated to occur during the test period. Peak day data are used in the allocation process to, among other things, determine the jurisdictional and nonjurisdictional portion of a company's sales. The Commission sometimes requires an average of three continuous days of maximum deliveries.

PGA   purchased gas adjustment. A special provision approved by the Commission allowing a company to make a special filing to increase its rates, without the usual suspension period, for the sole purpose of recovering currently the increases in its cost of purchased gas.

proper and adequate   Another subjective term, as in "proper and adequate depreciation rate." Means, in the final analysis, whatever the Commission and/or the courts say is "proper and adequate" on the basis of the facts in a given case.

public interest   Usually intended to mean the interest of the public generally, as opposed to the interest of any individual or company.

public utility   A business organization performing a service relating to or affecting all of the people in an area, usually under provisions of a franchise, charter, or certificate, and subject to special governmental regulations.

rate base   generally intended to mean the net number of dollars the company has invested at a given point in time in plant at "original cost," depreciated, and working capital required to operate the business and render "proper and adequate" service to the public. (FERC *Rules and Regulations* plus case history as expressed over the years by the Commission and the courts deal more specifically with what may and may not be included in rate base.)

rate of return   a percentage rate found to be "fair" which, when applied to the rate base, will provide the company with a "fair return" on its investment balanced between the interests of debt holders and stockholders.

rate zones   geographic areas of a company's operations, established to facilitate a design of rates that will properly reflect the cost of serving customers in different parts of the company's system.

reserve rates   rates designed for in-house use to calculate the number of dollars to be set up on the company's books each month as a reserve for possible refunds under a rate increase docket. (The cost of service supporting a rate increase filing is reduced on a judgment basis for any costs included, about which there may be serious doubt of the Commission's allowance. Such rates are applied to monthly sales volumes and demand billing units. The result is compared with actual customer billings for the month and the difference is set up as a reserve.)

return   the total dollars produced by application of the rate of return to the rate base, generally, interest on debt and the profit the company is allowed over and above the recovery of its operating expenses, depreciation, and taxes. A return *on* investment is distinguished from depreciation which is a return *of* investment.

rolled-in pricing   the accumulating into one single total cost figure of all gas purchased costs regardless of the varying prices/Mcf at which any purchases are made, and the dividing of this total cost by the total systemwide purchased volume. This will produce a systemwide average (or rolled-in) cost/Mcf.

SNG   synthetic natural gas, such as gas manufactured from petroleum products (for example, naptha), coal, or other substances.

SRO   sales refund obligation, a provision developed in settlement negotiations for a specific company, but not insisted upon by FERC staff in all settlements. It works this way: agreement is reached on the estimated annual sales volume used for determining a unit rate. Then, during the time such rates are in effect, if the actual annual volumes sold exceed the agreed-upon volume, the company must refund the fixed-cost component of the commodity revenue collected as a result of sales over the agreed-upon level. This refund may be offset by increases in jurisdictional rate.

stipulation and agreement   a document prepared to express in writing the agreement of the parties to a controverted matter such as a rate case. A stipulation and agreement settling all or part of a rate case must be submitted to the Commission for approval.

tariff   a published volume (or volumes) of effective rate schedules and general terms and conditions under which service will be rendered.

TBO   transport by others.

test period   A period of time extending 9 months beyond the end of the base period. Adjustment to base period data may be made for changes expected to occur during the test period, as provided in the FERC *Rules and Regulations.*

TFO   transport for others.

take or pay   contractual obligation of a buyer to pay for gas regardless of its ability to take delivery of the same.

tracker   special provision approved by the Commission, usually in the settlement of a rate case, which permits the company to adjust its rates without the usual suspension period to recover currently certain specified items of cost increases when incurred. The PGA is an example of a general tracker approved for all pipeline companies which apply for it. Another example of a generally approved tracker is for increases in research and development costs. Cost of gas transportation by others is yet another type of tracker, as is an SRO provision.

variable cost   Operating costs which vary either directly or indirectly in relation to any variation in the volume of gas sold and/or transported, e.g., compressor station fuel and expenses.

working capital   money necessarily invested in the business to carry on the day-to-day operations. For example: (1) 12½ percent of annual operation and maintenance expenses (excluding gas purchase cost and certain other items on which there is little or no lag in time between the incurring of expense and collection of revenue). It is generally conceded that there is a 45-day lag between the incurrence of most operating and maintenance expenses and the collection of related revenue. (2) Average monthly balances of: (1) materials and supplies (inventory), (2) prepayments (e.g., taxes, rents, insurance), (3) gas held in storage *for current use* (inventory), (4) advance payments on gas purchases, and (5) nonoperating bank balances. *Note:* The Commission does not always agree and allow all of the items listed.

Bibliography

1. Altari, A., and D. L. Klass (eds.), *Natural Gas Energy Measurement,* Elsevier Applied Science Publishers, London, and Institute of Gas Technology, Chicago, 1987.
2. Anderson, K. E., and B. D. Berger, "Gas Handling and Field Processing," Vol. 3, Plant Operations Training, PennWell Books, Tulsa, OK, 1980.
3. "Basic Natural Gas Transmission: A Self Study Course," prepared by Texas A&I University, Southern Gas Association, Dallas, TX 1989.
4. Birkhead, W. G., D. Mason, and M. P. Wilson, "Measurement Error Due to the Bending of Orifice Plates."
5. "Compressibility and Supercompressibility for Natural Gas and Other Hydrocarbon Gases," AGA Report No. 8, American Gas Association, Arlington, Virginia, 1985.
6. Datta-Barua, L., "Selection of Control Valve and Associated Instrumentation," *Proceedings,* International School of Hydrocarbon Measurements, Norman, OK, 1983.
7. Datta-Barua, L., "Systems Approach to Electronic Measurement Taken for Gas Custody Transfer," *Oil and Gas Journal,* Sept. 24, 1984.
8. Datta-Barua, L. and E. D. Woomer, "Electronic Measurement of Natural Gas," AGA Distribution/Transmission Conference, Chicago, 1986.
9. *Engineering Data Book,* Gas Processors and Suppliers Association.
10. Gas accounting course material, Southern Gas Association, Tulsa, OK, 4th edition, 1981.
11. *Gas Engineers' Handbook,* 1st ed., McGraw-Hill, New York, 1934.
12. Hale, D. (ed.), *Oil/Gas Pipelining Handbook,* Energy Communications, Inc., Dallas, TX, 1975.
13. "Manual for the Determination of Supercompressibility Factors for Natural Gas," PAR Research Project NX19, American Gas Association, Arlington, Virginia, December 1962.
14. "Measurement of Gas by Turbine Meters," AGA Report No. 7, American Gas Association, Arlington, Virginia, 1985.
15. "Orifice Metering of Natural Gas and Other Hydrocarbon Fluids," AGA Report No. 3, American Gas Association, Arlington, Virginia, 1985.
16. "Orifice Metering of Natural Gas and Other Hydrocarbon Fluids," AGA Report No. 3, Part 1, American Gas Association, Arlington, Virginia, 3rd Edition, Oct. 1990.

Index

About the Author

Lohit Datta-Barua, Ph.D., is a registered professional engineer in the state of Texas and superintendent of measurement at the United Texas Transmission Company in Houston. Previously, he was superintendent of measurement at the United Gas Pipe Line Company, and prior to that, was an instructor in civil engineering at the University of Houston. He also held the position of engineering/technical manager at American Science and Engineering Company. Dr. Datta-Barua is the author of *Digital Computer-Repair and Maintenance*, published by Tata McGraw-Hill. He received his M.S. in electrical engineering from the University of Houston and his Ph.D. in engineering from California Western University.